第一次下廚是七歲的時候，大家可能覺得很不可思議，那個時候三代一起移民到紐西蘭，一個屋子裡住著二十幾個人，父母與家裡的長輩人手不足的情況下，身為大孫女的我，只好跟在大人身邊一起下廚。

　　起初覺得料理很好玩，但由於家裡吃素，早期在國外取材不易，想吃什麼料理就必須自己做，12歲那年全家又移民到澳洲，又開始對於新的環境與新的食材重新了解，慢慢的我也開始四處取經，像是從認識的前輩學習與購買書籍，當時國外素食食譜並非盛行，所以我時常看葷食食譜做料理並將它改成素食，創造成新的菜色，沒想到不斷嘗試新菜色的成果，讓有吃過我的料理的人詢問我是不是能為他們開課，這也讓我更全心全意的投入烹飪美食的行列。

　　當我成年可以自己坐飛機四處旅行時，我發現原來有許多美食是國外吃不到的，我要如何把這些琳瑯滿目的美食帶回澳洲？那時我買了很多食譜回家研究，不斷想到底有什麼食材是可以取代的，什麼香料是在澳洲可以唾手可得的，或是當我回到我最初的家鄉台灣探親時，是否我也可以親手做我喜愛的外國料理不論是美式、印度、台灣……等，讓我都能親手做，這次透過書籍的傳遞我想把我所學從入門到進階傳授給大家，讓素食料理不只好吃也能結合各國美食並變化。

　　這是一本很幸福的書，很多人說美味的食物可以使人感到幸福，我希望這本書可以讓大家都感受到幸福的氛圍。

楊雅涵

楊雅涵 Ya-Han Sophie Yang

現任　Current role
　代課幼保老師、代課幼保廚師
　Childcare teacher, Childcare chef.

專長　Skills
　做料理、插花、幼保老師
　Cooking, Flower arranging, Childcare teacher.

Preface。

The first time I cooked was when I was seven. Many of you may think that is unbelievable. At the time, three generations of my family migrated to New Zealand and there were over twenty people living under the same roof. My parents and my uncles and aunties were often shorthanded hence, being the oldest granddaughter; I had no choice but to cook with the adults.

At the start, I thought cooking was fun but since our family were vegetarians it was difficult to obtain the ingredients overseas. Whatever we wanted to eat, we had to make it ourselves. When I was twelve years old, our family migrated to Australia. I grew an appreciation towards the new environment and new ingredients. Slowly, I began to explore different resources such as learning from different teachers and also studying books. At that time, there were not many vegetarian recipes overseas hence I often needed to read non-vegetarian recipes and adjust the recipe to become vegetarian and create new cuisines. I could not imagine that through experimenting with new foods made those that tried my cooking ask whether I can start a class to teach them. This made me more determined to fully devote myself in the culinary arts.

When I was old enough to travel on my own, I realized there were many delicacies not available in Australia. I wondered how could I bring these delicacies back to Australia? At that time, I brought back many recipe books with me to study and would constantly be thinking of which ingredients could be substituted and which condiments can be easily obtained in Australia. I would also think about when I returned to my home country to visit my relatives whether I am able to make my favourite exotic cuisine whether it is American, Indian or Taiwanese etc. on my own. Through this book, I wish to pass on to you what I have learnt starting from the very basics through to advanced levels of cooking. Vegetarian cuisine will not only taste good but can also be fused with foods from different cultures.

This is a book of happiness. Many say that great food can bring happiness to people. I hope this book can bring you happiness.

目錄 ▸ CONTENTS

1

基礎技巧

VEGETARIAN RECIPES
Basic Skills

材料介紹

Material Introduction

—

① 香椿

② 九層塔

③ 義大利香料（迷迭香）

④ 五香粉

⑤ 八角

⑥ 辣椒碎片或是胡椒子

⑦ 薄荷葉

⑧ 海苔素肉鬆

⑨ 肉桂條

⑩ 黑芝麻

⑪ 白芝麻

⑫ 義大利香料（奧勒岡葉）

⑬ 義大利香料（鼠尾草）

⑭ 義大利香料（巴西里 - 綠色、百里香 - 咖啡色）

⑮ 香菜

⑯ 甘草

---

① Toona

② Basil

③ Italian herbs (rosemary)

④ Five spice powder

⑤ Star anise

⑥ Chili flakes or pepper corns

⑦ Mint leaf

⑧ Seaweed floss

⑨ Cinnamon stick

⑩ Black sesame seeds

⑪ White sesame seeds

⑫ Italian herbs (oregano)

⑬ Italian herbs (sage)

⑭ Italian herbs (parsley-green, thyme-brown)

⑮ Coriander

⑯ Dried licorice root

① 百頁豆腐　　　　　⑮ 素蝦
② 海帶　　　　　　　⑯ 素丸子
③ 油豆腐　　　　　　⑰ 蒟蒻麵捲
④ 香菇羹素料　　　　⑱ 麵腸
⑤ 素魚　　　　　　　⑲ 菜脯
⑥ 豆腐　　　　　　　⑳ 雞蛋、皮蛋、鹹蛋
⑦ 海帶芽　　　　　　㉑ 洋菇
⑧ 麵筋球　　　　　　㉒ 杏鮑菇
⑨ 素肉塊　　　　　　㉓ 靈芝菇
⑩ 豆包　　　　　　　㉔ 黑香菇
⑪ 素魚　　　　　　　㉕ 黑木耳
⑫ 素料塊　　　　　　㉖ 髮菜
⑬ 素火腿　　　　　　㉗ 雪白菇
⑭ 素料塊　　　　　　㉘ 金針菇

---

① Louver bean curd
② Seaweed
③ Oily tofu
④ Mushroom chunk/ curd
⑤ Soy bean fish curd
⑥ Tofu
⑦ Dried kelp bud
⑧ Soy bean ball
⑨ Raw soy chunk
⑩ Mushroom chunk
⑪ Soy bean fish curd
⑫ Soy bean curd meat
⑬ Soybean-curd ham
⑭ Soy bean curd meat
⑮ Konjac shrimp
⑯ Soy bean curd ball
⑰ Konjac noodle roll
⑱ Gluten flour intestinal roll
⑲ Pickled dry radish
⑳ Egg, preserved century duck egg, marinate salty egg
㉑ Mushroom
㉒ King oyster mushroom
㉓ Marmoreal mushroom (brown beech mushroom)
㉔ Chinese black mushroom
㉕ Black fungus
㉖ Dried black moss
㉗ White beech mushroom
㉘ Enokitake mushroom

① 小白菜	⑧ 生菜	⑮ 豆芽菜	㉒ 茄子
② 玉米	⑨ 番茄	⑯ 三色豆	㉓ 紅蘿蔔
③ 芋頭	⑩ 綠花椰菜	⑰ 酸筍	㉔ 馬鈴薯
④ 蘆筍	⑪ 新鮮辣椒	⑱ 大白菜	㉕ 地瓜
⑤ 青椒	⑫ 南瓜	⑲ 菠菜	
⑥ 黃椒、紅椒	⑬ 薑	⑳ 白花椰菜	
⑦ 小黃瓜	⑭ 西洋芹	㉑ 高麗菜	

① Bok choy (Asian cabbage)	⑦ Cucumber	⑭ Celery	㉑ Cabbage
② Corn	⑧ Lettuce	⑮ Bean sprout	㉒ Eggplant
③ Taro	⑨ Tomato	⑯ Mix of frozen pea, corn and carrot	㉓ Carrot
④ Asparagus	⑩ Broccoli	⑰ Pickled bamboo shoot	㉔ Potato
⑤ Green capsicum	⑪ Fresh chili	⑱ Chinese cabbage	㉕ Sweet potato
⑥ Yellow capsicum, red capsicum	⑫ Pumpkin	⑲ Spinach	
	⑬ Ginger	⑳ Cauliflower	

① 鳳梨罐頭	⑥ 鮮奶	⑪ 奶油	⑯ 玉米粒
② 墨西哥辣椒	⑦ 優格	⑫ 花生	⑰ 大紅豆
③ 起司	⑧ 奶油乳酪	⑬ 油漬番茄乾	⑱ 鷹嘴豆
④ 番茄汁焗豆	⑨ 鮮奶油	⑭ 橄欖	
⑤ 巧克力醬	⑩ 哇沙米	⑮ 果醬	

① Pineapple tin	⑥ Milk	⑩ Wasabi	⑮ Jam
② Jalapeno chili	⑦ Greek yoghurt/plain yoghurt	⑪ Butter	⑯ Frozen corns
③ Cheese	⑧ Cream cheese	⑫ Peanut	⑰ Red beans
④ Bake beans	⑨ Thick cream or whip cream	⑬ Sundried tomato	⑱ Chickpeas
⑤ Chocolate jam		⑭ Olive	

① 番茄醬   ⑨ 調味好的辣椒粉   ⑰ 香菇味精

② 油   ⑩ 肉桂粉   ⑱ 辣椒油

③ 橄欖油   ⑪ 義大利香料   ⑲ 油辣椒

④ 素沙茶醬   ⑫ 醬油   ⑳ 鹽巴

⑤ 辣椒醬   ⑬ 烏醋   ㉑ 黑麻油

⑥ 芝麻香油   ⑭ 蜂蜜   ㉒ 糖

⑦ 辣椒粉   ⑮ 黑胡椒

⑧ 白胡椒   ⑯ 義大利巴撒米克香醋

---

① Tomato sauce   ⑨ Seasoned chili powder   ⑰ Dried mushrooms stock

② Oil   ⑩ Cinnamon   ⑱ Chili oil

③ Olive oil   ⑪ Italian herbs   ⑲ Chili paste

④ Vegetarian BBQ sauce   ⑫ Soy sauce   ⑳ Salt

⑤ Chili sauce   ⑬ Chinese black vinegar   ㉑ Black sesame oil

⑥ Sesame oil   ⑭ Honey   ㉒ Sugar

⑦ Chili powder   ⑮ Black pepper

⑧ White pepper   ⑯ Balsamic vinegar

① 米

② 迷你義大利麵

③ 麵包屑

④ 太白粉

⑤ 中筋麵粉

⑥ 海苔

⑦ 米粉

⑧ 蕎麥麵

⑨ 義大利麵

⑩ 麵筋粉

⑪ 吐司

⑫ 春捲皮、越南春捲皮

⑬ 細麵條

⑭ 冬粉

⑮ 杏仁

⑯ 腰果

⑰ 素肉燥

⑱ 杏仁片

⑲ 大黃豆

⑳ 松子

㉑ 美乃滋

㉒ 菲達起司

㉓ 酸奶油

㉔ 起司片

㉕ 墨西哥餅皮

---

① Rice

② Small pasta

③ Bread crumbles

④ Corn starch

⑤ Plain flour

⑥ Seaweed sheets

⑦ Rice noodles

⑧ Buckwheat noodles

⑨ Pasta noodles

⑩ Gluten flour

⑪ Toast

⑫ Spring roll pastry, Vietnamese spring roll wrapper

⑬ Thin noodles

⑭ Dried green bean noodles/glass noodles

⑮ Almond nuts

⑯ Cashew nuts

⑰ TVP (textured vegetable protein)

⑱ Almond slices

⑲ Soy beans

⑳ Pine nuts

㉑ Mayonnaise

㉒ Feta cheese

㉓ Sour cream

㉔ Cheese slices

㉕ Mexican tortilla

① 酪梨　　　　④ 香蕉　　　　⑦ 奇異果
② 綠蘋果　　　⑤ 鳳梨　　　　⑧ 檸檬
③ 紅蘋果　　　⑥ 草莓　　　　⑨ 柳橙

① Avocado　　　④ Banana　　　⑦ Kiwi fruit
② Green apple　⑤ Pineapple　⑧ Lemon
③ Red apple　　⑥ Strawberry　⑨ Orange

基本刀工
Basic Cutting Technique

—

❖ 削皮

1 手像拿筆一樣，拿穩削皮刀。由上往下削。就是由靠近自己的地方，向外削。

2 削到底的時候，可以轉一下，這樣就不會重複削到同一個地方。可以先削蔬菜的 ⅔，等到全部削好後，再削剛剛沒削到的 ⅓。

❖ 切片

刀子拿穩後，看好要切片的厚度，放好刀子的位子，然後一口氣切到底。持續這個動作，直到切完需要的材料或是切夠需要用的材料即可。

你在切的時候，如果你刀子拿的傾斜度越斜，你切出來的片會比較大，當然切的面也會比較大。可是如果你切的傾斜度比較小，切出來的面相對的也會比較小，主要是看個人的用途，還有想要呈現的效果。

切片動態影片 QRcode

❖ 切絲／切條

切絲／切條
動態影片 QRcode

使用剛剛切好片的食材，排整齊，切定不會滑動，一次不要重疊太多，確定要切絲或是條，來選擇粗細度。通常切片和切絲的厚度相同。

如果想要一口氣切完，可以用斜疊的方式（每一片食材只重疊一半，這樣比較不會滑動），滑動幅度比較不會太大，切的時候比較安全。

❖ 切丁／切塊

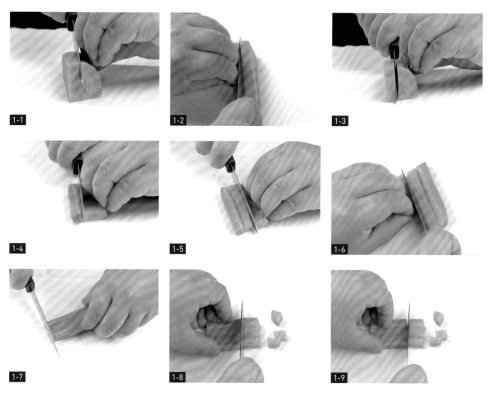

刀子拿穩後，看好自己要切片的厚度，放好刀子的位子，然後一口氣切到底。持續這個動作，直到切完需要的材料或是切夠需要用的材料即可。

在切的時候，要比切片的厚度寬一點，待會切條時才會一樣寬，到切成丁的時候，才會變成很均勻的丁，每一面差不多大小。

切丁／切塊
動態影片 QRcode

❖ 切末

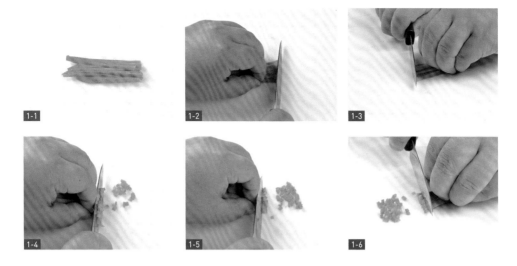

1-1 1-2 1-3 1-4 1-5 1-6

剛剛切好的細絲，放整齊，同一端一樣擺整齊，在切的時候比較容易均勻。切末和剛切細絲的寬度相同，切出來時有一點像小丁。

切末動態影片 QRcode

❖ 切細末

1-1

1-2

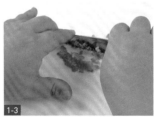

1-3

當切好末時，還想要更細，可以把材料集中起來，把刀子放在中間，用按住食材的那隻手，按住刀子一端，然後握住刀子的那隻手，上下切，由左邊切到右邊，再從右邊切到左邊。這樣來回切，食材就會越來越小，慢慢的變成了細末。

切細末
動態影片 QRcode

❖ 滾刀

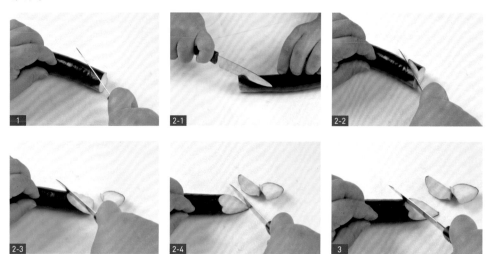

滾刀就是你在切食材的時候，以 90 度角一邊切食材一邊滾動食材。

滾刀動態影片 QRcode

1　看好你要下刀的地方，確定你要的食材大小，以 45 度角切下去。

2　將按住食材的那隻手，把切好的食材往自己站的方向滾一點，直到剛剛切好那面食材的切面處朝上方。然後再切過那個面的一半的位子，看好你要下刀的地方，一樣以 45 度角切下去。

3　一直重複持續這樣的動作，直到切完食材。

👨‍🍳 備註

◆ 在切的時候，拿刀子的手要拿穩、拿順。在輔助的另外一隻手（拿菜的那隻手）。要用手指頭穩菜，不要讓菜滾動，如果會滾動，很容易切到手。在切菜的時候，按住菜的那隻手，手指頭要縮起來，用手指中間的關節頂住刀子的邊緣，這樣就不會傷到手。

◆ 在切菜的時候，重點是不要讓菜搖動，如果是會滾的菜，可以先切到一邊再切另一邊，這樣你翻過來時，剛切好的平面那一面朝下，這樣菜就比較不會因滾動造成危險。

Basic Cutting Technique

❖ Peeling

01   Hold the peeler like holding a pen, or how you find it comfortable, start from the top and peel your way downwards. This means you start at the end closer to you and peel it away from you.

02   When you reach to the bottom of the vegetable, you turn it slightly, this will help you peel all the round the vegetable without peeling the same place you have peeled. You can also peel $\frac{2}{3}$ of the vegetable, when you finish peeling, you flip over, and peel the rest of the $\frac{1}{3}$ of the vegetable.

❖ Slice

Hold the knife firmly, look at the thickness you would like, place the knife on the vegetable, cut it all the way down in one go. Continue this process until you have finish cutting the whole vegetable, or until you have enough slices you need.

When you are making the cut, the more angle you make the cut, the larger piece of the slice it will turn out, of course, than you will have a larger facing. But the smaller angle you make each cut, the slice will be smaller, and the facing will be smaller. It comes down to how you like your slice, and what size you like it, also what kind of presentation you would like your dish to be presented.

❖ Shred/Julienne

Start the with slice cutting for the vegetable you need, stack the already cut sliced vegetable, make sure you don't stack to many slices on top of each other, as this can make it very hard to control, and hold the vegetable without vegetable sliding away. Look at the thickness you would like for the shred/julienne, usually this is the same thickness as when you were cutting the slice.

If you would like to cut everything in one go, you can stack the vegetable by over lapping each piece by half of each slice, as this will give you more control, and safer making each cut.

❖ Cubes

Hold the knife firmly, look at the thickness of the vegetable you would like to cut, place the knife on the vegetable, and cut it to the bottom of the vegetable in one go. Continue with this process till the vegetable has all been cut, or until you have enough quantly of the vegetable you need.

When you cut the vegetable into slice, you need to know the size of the cubes you would like, so when you cut into the slice, than into shreds/julienne it is all the same thickness, so when you cut it into cubes, it is also same thickness, which will turn into cubes.

❖ Dice

Use the precut shreds/juliennes, place them together in a small bunch, make sure one end of the shreds/juliennes are even, this will help make the dicing much easier to control. Cut the dice around same thickness as when you were cutting the shreds/julienne. Cut it all the way until finish.

❖ Fine Dice

This is when you finish cutting the dice, and you would like to cut it more finely, you can gather the dice into a small pile, than place the knife in the middle of the pile, hold the end down with one hand, and move the hand that is holding the knife with the other hand, move the knife up and down, repeating left to right, than right to left, until you have cut the vegetable into fine dice how you would like it to be.

❖ Roll Cut

This roll cut means when you are cutting the vegetable, as you made each cut, you also turn the vegetable around 90degree, on way.

01   Look at the place you would like to make the cut on the vegetable, depends on the size you like, on a 45degree angle, you make your first cut.

02   The hand you press/hold on the vegetable, after you made your cut, turn/roll the vegetable toward you, so the cut you made (the face of the cut) is facing the top, around half way of the face, look at where you going to make the next cut, same, at 45degree angle make your second cut.

03   Continue the process until all is finish, or until you have enough of the cut vegetables you need.

 NOTE

- When you are making each cut, you need to hold the knife firmly, and comfortable. Use your other hand to support the vegetable. Make sure when you are supporting the vegetables, you use your fingers to support it, and don't let the vegetable roll around, because if you rolls, you can easily cut your finger/hand. When you are cutting, the hand you are holding the vegetable, make sure you have tucked all your fingers close to your palm, and you are using your middle knuckle to stabilize the knife edge/knife side. This will support you, so you don't cut your hand or fingers during the cutting process.

- When cutting the main aim is don't let the vegetable you are cutting roll around, if it's the type of vegetable that will roll around, you can make a thin cut on a side, and than flip that side that you made the cut facing the bottom (the chopping board), so it sit solidly on the chopping board not able to roll around, this will make your cutting safer and much easier to control.

BASIC
3

基礎技巧

高湯
Vegetable Stock / Broth
—

◆ **材 料**

紅蘿蔔	1 ～ 2 條	甘蔗汁	1 ～ 2 杯
玉米	2 ～ 4 條	水	6 ～ 10 杯
西洋芹	½ 把	鹽巴	1 ～ 2 大匙
高麗菜	½ ～ 1 顆		

◆ INGREDIENT

Carrots	1-2 whole	Sugar cane juice	1-2 cups
Corns	2-4 whole	Water	6-10 cups
Celery	½ bundle	Salt	1-2T
Cabbage	½-1 whole		

◆ 做法

1　洗，削後，將紅蘿蔔、玉米、西洋芹、高麗菜切大丁或大塊。準備所有材料。

2　將所有材料放置一個大鍋子中。

3　加入水、甘蔗汁。

4　把所有材料煮至滾，然後關至小火慢慢熬約 1 ～ 1.5 個小時。完成後即可以使用。
（註：在高湯中的蔬菜可以濾出來，或放在高湯中，主要是看高湯的用途。）

5　可以加入鹽巴調味。

🍳 備註

◆ 高湯可以用任何蔬菜煮，也可以加入一些素料一起熬煮。

◆ 高湯可以用以下的材料一起煮：

1. 海帶、黑香菇、醬油、味噌、糖跟鹽巴。

2. 黑香菇、大白菜、紅蘿蔔、玉米、小黃瓜跟香菜。

3. 任何蔬菜、黑香菇、甘草、八角、肉桂條、五香粉、醬油、糖跟鹽巴（胡椒）。

4. 蘋果、鳳梨、任何蔬菜、鹽巴跟糖。

※白蘿蔔、番茄、地瓜、南瓜、冬瓜、紅菜頭等都是可以用來煮高湯的蔬菜。

Vegetable Stock/Broth

◆ METHOD

01 Wash, peel and cut carrots, corns, celery, cabbage into large chunks. Prepare all ingredients.

02 Place everything inside a large pot.

03 Add water into the pot, and add in the sugar cane juice.

04 Bring everything to boil, and then simmer for 1-1.5hours. Than it is ready. (The vegetables can be drain out, or keep inside the soup, depends what the stock is used for).

05 Can add some salt to season if desire.

🍳 NOTE

- You can use any kind of vegetables to cook the vegetable stock/broth, you can even add some soy bean product/meats to cook together to give flavor.

- You can add the below ingredients to cook the vegetable stock/broth:

  1. Seaweed, Chinese black mushroom, soy sauce, miso paste, sugar and salt.

  2. Chinese black mushroom, Chinese cabbage, carrots, corns, cucumbers and corianders.

  3. Any kind of vegetables with Chinese black mushroom, dried licorice roots, star anise, cinnamon stick, five spice powder, soy sauce, sugar and salt (pepper).

  4. Apple, pineapple, vegetables, salt and sugar.

  * White radish, tomato, sweet potato, pumpkin, winter melon, beetroot etc. can all be use as one of the type of vegetable cooking vegetable stock/broth.

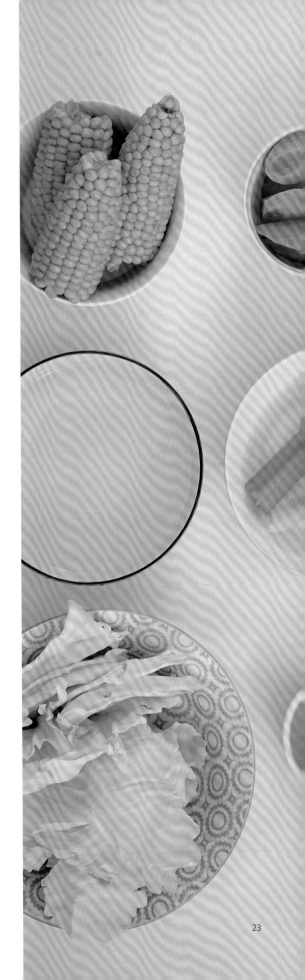

CHAPTER

2

醬
汁

VEGETARIAN RECIPES

Sauces

Recipe 01

萬用番茄醬

Universal Tomato Sauce

◆ 材料

番茄（切丁）	4 杯	鹽巴	1 大匙
番茄（泥）	2 杯	糖	2～3 大匙
番茄糊	½ 杯	胡椒	½～1 大匙
水或高湯	¼ 杯	義大利香料	1～2 大匙
油	4～6 大匙		

◆ 工具

果汁機

◆ 做法

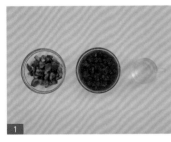

1　先將番茄洗好，切丁約 1.5×1.5 公分，切滿兩杯後放置旁邊備用。

2　剩下的番茄加入⅓的水或高湯，放入果汁機打成泥狀。

3　混合成泥狀。

4　熱鍋後加入油。

5　待油熱後（鍋內冒出煙），或是用筷子插入油鍋內測試油溫，當筷子邊緣有冒出小泡泡，代表油溫已夠熱。

6　放入切丁的番茄。

7　炒至小滾。

8　再加入番茄泥。

9　待鍋內食材大滾。

10　加入番茄糊。

11　加入鹽巴、糖、胡椒及義大利香料。

12　讓食材滾 15 ～ 30 分鐘，確認食材變濃稠狀後關火，讓食材冷卻後，即可放入保鮮盒，放置冰箱冷藏或冷凍保存。

☙ 備註

◆ 這個醬可以用於：披薩底層的醬、義大利麵、千層麵、羅宋湯。

◆ 如果你不想要湯內有塊狀，就把全部的番茄打成泥。

◆ 味道依個人喜好再做調味。

◆ 將每次需要用的量做單次包裝，這樣食材比較容易保存。

Universal Tomato Sauce

◆ INGREDIENT

Tomatoes	4 cups	Oil	4-6T
Tomato puree	2 cups	Salt	1T
Tomato paste	½ cup	Sugar	2-3T
Water or vegetable stock/ broth	¼ cup	Pepper	½-1T
		Italian herbs	1-2T

◆ TOOL

Blender

◆ METHOD

01 Diced tomatoes: wash the tomatoes, diced tomatoes 1.5cmX1.5cm up to 2cups.

02 Puree tomatoes: rest of the tomatoes adds ⅓ cup of water or vegetable stock/broth, mix in the blender.

03 Mix it into puree.

04 Add oil to pan.

05 Wait till it is heated with small amount of smoke coming from the oil inside the pan. Or use chopstick to check the oil, when side of chopstick makes little bubbles, it is hot enough.

06 Add the diced tomatoes.

07 Fried till boiled.

08 Than add the puree tomatoes.

09 Wait till boiled.

10 Add the tomato paste.

11 Add salt, sugar, pepper and Italian herbs.

12 Let it simmer for 15minutes-30minutes, until it has thickened. Than you can rest it to cool, to store in dried containers, fridge or freezer.

🍲 NOTE

◆ This sauce can be used for pizza base sauce, pasta, lasagna, Italian soup.

◆ If you don't want any chunky bits in the sauce, just puree everything.

◆ Flavoring can be adjust to own taste.

◆ Freeze product in small portion, only take out the amount you need.

醬汁 SAUCES

Recipe 02

披薩底醬

Pizza Base Sauce

白醬
Cream Sauce

◆ 材料

麵粉	12～14大匙	鹽巴	少許
鮮奶	5～6杯	胡椒	少許
奶油	1～2大匙		

◆ 工具

攪拌器

◆ 做法

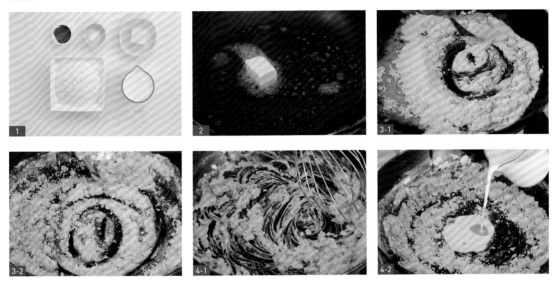

1　備好所有材料。

2　熱鍋後，調至小火，加入奶油。

3　待奶油融化後，慢慢加入麵粉，並一
　　邊攪拌。

4　使用攪拌器攪拌均勻，再分次加入鮮
　　奶煮至濃稠，加入鹽巴、胡椒。

5　混合均勻，煮至沸騰，即可關火備用。

🧑‍🍳 備註

◆ 在煮的過程中，要不斷的攪拌，不然易黏鍋或有烤焦的情況發生。

◆ 如果鍋中有結塊，可以倒至碗中，用電動攪拌器攪拌均勻後再倒回鍋子，煮至滾。

◆ 當我們製作一般白醬時，牛奶不用加很多，煮好時會是膏狀，可是如果你要煮成義
　大麵的白醬，鮮奶要加比較多，讓它變成滑順又濃稠的狀況，有一點像濃湯。

◆ 你也可以偷吃步，就是你先把麵粉與一些鮮奶混合均勻，變成有一點稀的麵糊，然
　後把剩下的鮮奶在另外的鍋子裡加熱煮滾，加入調味料，當鍋子裡面的鮮奶已經滾，
　這個時候你一邊攪拌，一邊加入剛剛拌好的麵糊。當你的白醬已經煮到你要的濃稠
　度時，就可以停下。剩下的麵糊就不要再用了。可是如果麵糊不夠，這時候你要關火，
　再準備需要的麵糊，準備好時，再開火繼續加入，將白醬拌至你要的濃稠度。

- Pizza Base Sauce -

Cream Sauce

◆ INGREDIENT

Plain flour	12-14T
Milk	5-6 cups
Butter	1-2T
Salt	some
Pepper	some

◆ TOOL

Whisk

◆ METHOD

01   Prepare all ingredients.

02   Heat pan, use low heat, add butter.

03   When butter has melt, slowly add the plain flour, and need to continue to stir.

04   Mix well using a whisk, and then add milk, gradually until the mix is smooth and creamy, than add salt and pepper.

05   Bring it to boil.

☐ NOTE

- Need to continue to stir while it is cooking, so it doesn't get burnt or stick on the bottom of the pan.

- If it is really lumpy, you can pour out the mixture into a bowl, use an electric whisk to mix it well, and then pour it back into pan to bring it to boil.

- When you are making this, because it's going to a paste, you don't need to add as much milk as how you normal make white sauce. But if you are making pasta, than you would need to add more milk to make it thick and smooth/runny more like a thick soup feel.

- You can do it cheating way, that is mix the plain flour in some milk, until it becomes a runny smooth paste, than add rest of the milk in a pot, bring it to boil, add all your seasonings, taste to check it is how you like it, when the milk are boiled, than you slowly add in the plain flour mixture, stir as you add it slowly, stop when it becomes the type of consistency you want. If you want it thicker, and you run out of the plain flour mixture, turn the heat off and make some more. But if it's enough, than you stop even if you have some more of the mixture left.

青醬

Pesto Sauce

◆ 材料

九層塔 .... 2 ～ 3 杯

油 .... ½ ～ ⅔ 杯

松子 .... ½ 杯

◆ 工具

果汁機

◆ 做法

1　將九層塔洗淨、晾乾或擦乾，放旁備用。

2　用一個乾的鍋子，熱鍋後加入松子。

3　用小火加入松子炒至金黃色。

4　將九層塔放入果汁機。

5　加入油，與九層塔打至均勻。

6　加入松子，與九層塔打至均勻。

7　將打好的青醬放置罐子內，保存至冰箱冷藏。

🍳 備註

◆ 如果九層塔或是瓶子沒有用乾，食物會比較容易壞掉，或是發霉。

- Pizza Base Sauce -

Pesto Sauce

◆ INGREDIENT

Basil	2-3 cups
Oil	½ - ⅔ cup
Pine nuts	½ cup

◆ TOOL

Blender

◆ METHOD

01　Wash, drain and dry the basil, set aside ready to use.

02　Heat the pan, place pine nuts into the pan.

03　Use low heat to stir fried the pine nuts until golden.

04　Add basil into the blender.

05　Add oil into the blender, mix with the basil.

06　Blend the pine nuts, into the basil mixture. Mix well.

07　Store in a tight jar, place in refrigerator.

🍳 NOTE

If the basil or the jar is not dry properly, it can easily go off, or have mold.

哇沙米、美乃滋、海苔絲
Wasabi, Mayonnaise, Seaweed Shreds

◆ 材 料

A

麵粉	3～4大匙
鮮奶	1杯
奶油	½大匙
鹽巴	少許

B

哇沙米	1～2小匙
美乃滋	1～2大匙
海苔絲	少許
芝麻（炒過或烤過的）	少許
胡椒	少許

◆ 工 具

攪拌器

◆ 做法

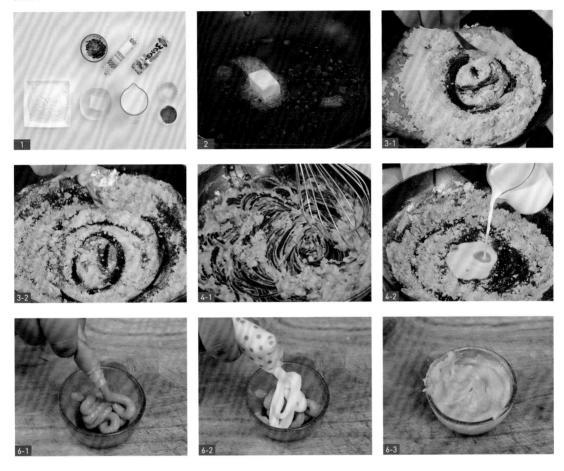

1　備好所有材料。

2　熱鍋後，調至小火，加入奶油。

3　待奶油融化後，慢慢加入麵粉，並一邊攪拌。

4　使用攪拌器攪拌均勻，再分次加入鮮奶煮至濃稠，加入鹽巴、胡椒。

5　混合均勻，煮至沸騰，即可關火備用。

6　充分混合美乃滋與哇沙米，放旁備用。

7　A 食材用於披薩的底層，並加入其他食材、起司，可放入烤箱做烘焙的部分；B 醬料作
　　為披薩的頂層，使用於披薩烤好出爐後，淋在最上面，最後放上海苔絲及芝麻，即可
　　享用。

Wasabi, Mayonnaise, Seaweed Shreds

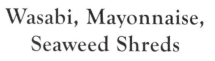

♦ INGREDIENT

A

Plain flour	3-4T
Milk	1 cup
Butter	½T
Salt	some

B

Wasabi	1-2t
Mayonnaise	1-2T
Seaweed shreds	some
Roasted sesame	some
Pepper	some

♦ TOOL

Whisk

♦ METHOD

01 Prepare all ingredients.

02 Heat pan, use low heat, add butter.

03 When butter has melt, slowly add the plain flour, and need to continue to stir.

04 Mix well using a whisk, and then add milk gradually until the mix is smooth and creamy, than add salt and pepper.

05 Bring it to boil.

06 In a bowl, mix wasabi and mayonnaise, set aside ready to use.

07 Mixture A (white sauce) is use as pizza base sauce, where it can be place in oven to bake with other ingredient and cheese, and B mixture is use when pizza has finish baking, where the sauce is add on top than the seaweed shreds and roasted sesame, just before serving.

醬汁 SAUCES

Recipe 03

酪梨醬

Avocado Dip

備註

◆ 若想要濃稠點，可再多加入奶油乳酪及油。

◆ 若不喜歡奶油乳酪或酸奶油，可用美乃滋代替。

酪梨醬

◆ 材料

酪梨	5～8 顆	油	少許
酸奶油	5～7 大匙	鹽巴	少許
奶油乳酪	8～10 大匙	胡椒	少許
檸檬汁	2～3 大匙		

◆ 工具

果汁機

◆ 做法

1　用刀子，將酪梨切半，將籽取出，也將皮剝掉後，將酪梨放入碗中。準備所有材料。

2　將剝好的酪梨放在果汁機裡面。

3　加入檸檬汁。還有少許的油。

4　加入奶油乳酪。

5　加入酸奶油、鹽巴、胡椒至果汁機裡面。

6　將所有食材均勻打成濃稠狀，試味道。如果你喜歡乳酪多一點，你可以加多一點奶油乳酪，如果你喜歡酸一點的，你可以加入多一點檸檬汁或是酸奶油。放在冰箱保存直到準備使用再拿出來。

Avocado Dip

◆ INGREDIENT

Avocados	5-8 whole
Sour cream	5-7T
Cream cheese	8-10T
Lemon juice/lime juice	2-3T
Oil	some
Salt	some
Pepper	some

◆ TOOL

Blender

◆ METHOD

01   Use a knife, cut the avocados in half, remove the seed, than remove the skin or you can just peel the avocados; remove the seed, place avocados in a bowl. Prepare all ingredients.

02   Place peeled avocados in a blender.

03   Add in the lemon juice, and a bit of oil into the blender.

04   Add cream cheese into the mixture.

05   Add sour cream, salt and pepper into the blender.

06   Blend it all together, until it becomes a smoother thick paste. Taste to make sure it is to your liking, if you like more cheese flavor you can add more cream cheese, if you like more sour flavor you can add more lemon juice or sour cream. Place in the refrigerator until ready to serve.

🍴 NOTE

- If you like more richness, add a bit of cream cheese and oil.

- If you don't like sour cream or cream cheese, can you mayonnaise instead.

Recipe 04

墨西哥沙沙醬

Salsa Dip

◆ 材料

番茄	5～8 顆	鹽巴	少許
青椒	½ 顆	胡椒	少許
黃椒	½ 顆	醬油	少許
玉米粒	½ 杯	香菇醬油膏	少許
西洋芹	½ 杯	高湯	1～2½ 杯
墨西哥辣椒	¼～½ 杯	檸檬汁	2 大匙
萬用番茄醬	4～6 大匙	油	2 大匙

◆ 做法

1　將所有蔬菜洗淨後，將番茄、青椒、黃椒、墨西哥辣椒、西洋芹切小丁。準備所有食材。

2　熱鍋後加油，加入所有食材。

3　加入萬用番茄醬，或是從頭做起，將新鮮的番茄丁跟番茄泥放在鍋中一起煮。（註：萬用番茄醬請參考 P.27。）

4 　加入檸檬汁。

5 　加入高湯，蓋過所有食材。

6 　煮滾。

7 　加入鹽巴、胡椒、醬油、香菇醬油膏。（註：如果喜歡可以額外加入新鮮辣椒。）

8 　繼續滾，試味道看是不是自己喜歡的味道後，轉小火滾15～30分鐘。直到醬汁變少，還有變濃稠點。

9 　熱食或冷食都可食用，也可搭配餅乾或麵包風味更佳。

備註

◆ 可依個人喜好做調味。

◆ 若你不喜歡吃煮過的莎莎醬也可食用生的，風味會更清新。做法為：將青椒、黃椒、玉米、香菜切丁後，加入墨西哥辣椒、鹽巴、胡椒、攪拌均勻後（也可依個人喜好加入芝麻香油、檸檬汁、新鮮辣椒），即可享用。

◆ 如果想加香菜，需等食材冷卻後再放，風味會更清新、爽口。

Salsa Dip

◆ INGREDIENT

Tomatoes	5-8 whole	Salt	some
Green capsicum	½ whole	Pepper	some
Yellow capsicum	½ whole	Soy sauce	some
Frozen corns	½ cup	Mushroom soy paste	some
Celery	½ cup	Vegetable stock/broth	1-2 ½ cups
Jalapeno chili	¼- ½ cup	Lemon juice	2T
Universal tomato sauce	4-6T	Oil	2T

◆ METHOD

01 Wash and diced tomatoes, green and yellow capsicum and celery into small cubes. Mince diced jalapeno chili. Prepare all ingredients.

02 Heat the pot, add oil, and place all ingredients in.

03 Add the homemade universal tomato sauce, or you can make it from start, with more tomatoes and tomato puree, with the rest of the ingredients. (Please refer to P.27 for universal tomato sauce recipe.)

04 Add the lemon juice.

05 Add the vegetable stock/broth; just cover all the ingredients in the pot.

06 Bring it all to boil.

07 Add salt, pepper, soy sauce, mushroom soy paste. (Can add fresh chili if desire.)

08 Keep it boiling, taste it, if it's how you like it, than let it simmer for 15-30minutes. Until sauce has reduce, and it becomes a bit thick.

09 Serve it hot or cold. With crackers or toasted bread.

☞ NOTE

◆ Flavoring can change to own desire.

◆ You don't have to cook it if you like more of the fresh and raw taste. In this case you only need: dice tomatoes, capsicums, frozen corns, coriander and jalapeno chili with salt and pepper, mix together (can also add bit of olive oil or sesame oil, lemon juice or fresh chili if you like it hot.), ready to serve.

◆ Add in the coriander last, after everything has cooled down, it will give the dish more freshness flavoring.

烤合椒醬

Capsicum Dip

備註

◆ 甜椒、青椒可以整顆烤，也可以剖半後烤，只是烤的時間長短不同。烤好的甜椒、青椒，如果是整顆或是剖半，甜椒、青椒的皮跟肉開始分離後即可取出放涼，用手將甜椒、青椒皮剝起、丟棄。但如果覺得麻煩，不一定要將皮剝掉。食材可依個人喜好做調整。

◆ 若不喜歡奶油乳酪或是酸奶油，可用其他蔬菜代替（蔬菜可跟甜椒、青椒一起放入烤箱烤，待涼後，放入果汁機一起攪拌）。

◆ 香菜與檸檬汁，可依個人喜好調整，增加清新的風味。你喜歡當然也可以加入義大利巴撒米克香醋代替檸檬汁。

◆ 你也可以加入一點辣，像是新鮮辣椒、辣椒粉、墨西哥辣椒等。

◆ 你可以加入九層塔、巴西里、薄荷葉。

◆ 你也可以加入烤過的豆類、堅果。

◆ 也可以加入優格，給它更濃郁的口感。

◆ 材料

各色甜椒、青椒
..............................8 ～ 15 個
油.......................2 ～ 3 大匙
香菜.........................少許

檸檬汁..............1 ～ 3 大匙
鹽...........................少許
胡椒.........................少許

◆ 工具

烤盤
烤盤紙
果汁機

◆ 做法

1　將甜椒、青椒洗淨、切半，將裡面的籽都去掉，然後切成條狀，放旁備用。
　　將烤盤紙放置烤盤上，準備所有食材。

2　加入油至青椒中。

3　混合均勻，確定每一個甜椒、青椒都有包覆上一成薄薄的油後，放在烤盤上，
　　不要重疊。熱烤箱 170 ～ 200℃，將烤盤放入烤箱烤 10 ～ 20 分鐘，或是烤
　　至甜椒、青椒變軟或是變色。

4　將烤好的甜椒、青椒拿出烤箱，放涼。

5　將烤好的甜椒、青椒放入果汁機。

6　加入香菜、檸檬汁、油、鹽巴跟胡椒，一起攪拌均勻成膏狀，即可使用。

Capsicum Dip

◆ INGREDIENT

Mix color capsicums
.... 8-15 whole

Oil 2-3T

Corianders .... some

Lemon juice 1-3T

Salt .... some

Pepper .... some

◆ TOOL

Baking tray

Baking paper

Blender

◆ METHOD

01 Wash, and cut the mix color capsicums in half, remove the seeds inside, then cut it all into thick shreds/julienne. Set aside ready to use. Place a sheet of baking paper on a baking tray. Prepare all ingredients.

02 Add oil into the mix color capsicums.

03 Mix it well, making sure every piece of the mix color capsicums have been coated with oil. Then place it on baking tray. Making sure it's only a single layer, the mix color capsicums are not on top of each other. Heat up the oven at 170-200degree, and roast the mix color capsicums for 10-20minutes or until the mix color capsicums have become soft and has colored.

04 Take out of the oven. Let it cool.

05 Place the roasted capsicums in a blender.

06 Add corianders, lemon juice, oil, salt and pepper into the mixture. Blender it all together until it becomes a thick paste. Ready to serve.

🍳 NOTE

• You can bake the mix color capsicums as a whole or cut it in half, when the mix color capsicums are ready, the skin and the flesh will fall apart slightly, than they are ready to take out, let mix color capsicums cool, then pull off the mix color capsicum skin, and use only the flesh, but if you can't be bother to separate it two, then you don't need to do this step. The ingredient can be alternate to suit each individual.

• If you don't like to add cream cheese or sour cream, you can always just have roasted mix color capsicums and other vegetable ingredient mixed together (You can roast the vegetables with the mix color capsicums together. Than you place it all in blender to blend.).

• Corianders and lemon juice is to give it a fresh taste, you can add as you desire. You can even add some balsamic vinegar to the mixture instead of using the lemon juice.

• If you want to give it a little spicy, you can add fresh chili, chili flakes or jalapeno chili to the mixture.

• Basil, parsley or mint leaves can be added if desire.

• Can add some roasted nuts.

• Greek yoghurt/plain yoghurt can be add to the mixture too, to give it more creamy flavors.

素肉燥

Vegetarian Mince

 備註

◆ 加入香菇水是為了讓料理的香味更重。

◆ 若加入過多的水，湯汁收乾的時間會拉長，且有可能因食材吸飽湯汁而收不乾。但如果有人喜歡將湯汁淋在飯上面吃，也可以刻意加多一點的水。

◆ 可以加入鹽巴，胡椒或是新鮮辣椒調味。

◆ 若沒吃完時，可待冷卻後，放入冷藏可保存 3 ～ 5 天，冷凍可保存 6 個月。

素肉燥

◆ 材料

素肉燥	4 杯	糖	½ ～ 1 杯
黑香菇	6 ～ 10 朵	香菇水	½ ～ 1 杯
黑木耳	4 ～ 6 朵	水	1 ～ 2 杯
香椿	1½ 杯	油	1 ～ 1½ 杯
醬油	1 ～ 1½ 杯		

◆ 做法

1 用水泡軟素肉燥以及黑香菇，濾乾素肉燥，用手擠出素肉燥內的水分，放旁備用。將黑香菇清洗乾淨，用手擠出黑香菇內的水分，並且保留泡黑香菇的水，放旁備用。將香椿、黑香菇、黑木耳，切成小碎塊或是用果汁機分別打碎。放旁備用。準備所有食材。

2 熱鍋後加油。關小火。

3 放入黑香菇爆香，炒至金黃色。

4 再放入黑木耳，翻炒約 2 分鐘。

5 再放入素肉燥，繼續翻炒約 5 ～ 10 分鐘。

6 再加入香椿，用中小火繼續翻炒，將食材都混合均勻。

7 加入醬油、糖。

8 這時將火轉為大火，加入香菇水，如果有需要可以再加入少許的水，加至水分快要蓋過所有食材即可。翻炒至水分被收乾。放旁備用。

Vegetarian Mince

◆ INGREDIENT

Textured vegetable protein (TVP)	4 cups	Sugar	½-1 cup
Chinese black mushrooms	6-10 whole	Mushroom water	½-1 cup
Black fungus	4-6 whole	Water	1-2 cups
Toona	1 ½ cups	Oil	1-1 ½ cups
Soy sauce	1-1 ½ cups		

◆ METHOD

01 Soak the TVP and Chinese black mushrooms separately until soft, drain TVP, use hand to squeeze the excess water out of TVP, set aside ready to use. Wash the Chinese black mushrooms and use hand to squeeze the excess water out for Chinese black mushrooms, keep the mushroom water that was soaking the Chinese black mushrooms for later use. Finely chop or use a blender to separately blend toona, Chinese black mushrooms and black fungus. Set aside ready to use. Prepare all ingredients.

02 Heat a pan, add oil. Cook on low heat.

03 Add in Chinese black mushrooms, stir fried till golden.

04 Then add black fungus. Stir fried it for 2minutes.

05 Then add TVP, continue to stir fried for 5-10minutes.

06 Add in the toona. Continue to cook it on low-medium heat. Till it is all mixed.

07 Then add sugar, soy sauce.

08 Turn to high heat and then add the mushroom water, and some more water if required. As long as it is nearly covering all the ingredients. Stir fried till the moist is all absorbed. Ready to serve.

🍳 NOTE

- Reason for adding the mushroom water is to give the dish a stronger flavor.
- If you add to much water, your cooking time will be longer, and if you add to much water, because the ingredient has soak up enough water, you will left with a lot of juice left. Of course, if you like to have lot of the sauce to go with rice, it is okay as well.
- Can add salt, pepper or fresh chili if desire.
- If there are left overs, let it cool, and store in a container. This can be store in the fridge for 3-5days or freezer up to 6months.

前

菜

VEGETARIAN RECIPES

Appetizer

醃黃瓜

Soy Pickle Cucumber

◆ 材 料

醬油	70 毫升	二號砂糖	74 克
白醋	50 毫升	小黃瓜	600 克
甘草	5 ～ 10 片		

◆ 做法

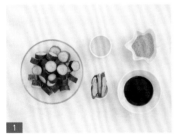

1　將小黃瓜洗淨，擦乾，切成厚塊，約 1 ～ 2 公分厚。準備好所有食材。

2　在鍋中加入醬油。

3　加入白醋。

4　加入甘草。

5　加入二號砂糖。

6　一邊攪拌醬汁，並煮至滾。

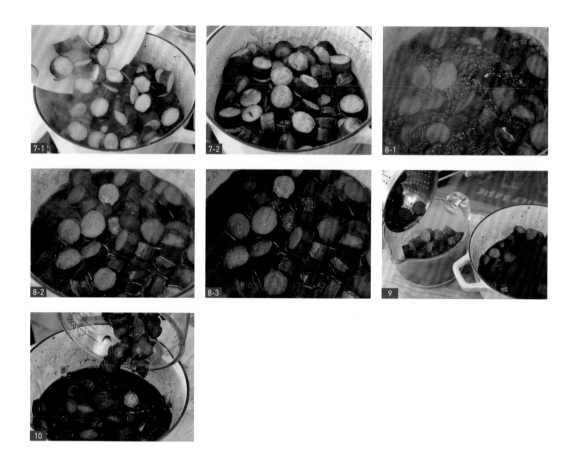

7 加入小黃瓜至醬中，開小火煮約 10 分鐘。

8 攪拌至所有小黃瓜都上色且煮滾。

9 將醬汁與小黃瓜分開後，將兩個都放涼。

10 再將放涼的小黃瓜與醬汁混合後，再放入一個乾淨的瓶子中，保存在冰箱，即可享用；可保存 1 星期。

Soy Pickle Cucumber

◆ INGREDIENT

Soy sauce	70cc	Raw brown sugar	74g
White vinegar	50cc	Cucumbers	600g
Dried licorice roots	5-10 slice		

◆ METHOD

01 Wash cucumbers, dry it, and then cut into thick slice of 1cm-2cm. Prepare and measure all ingredients.

02 Add soy sauce into a pot.

03 Add in white vinegar.

04 Add dried licorice roots.

05 Add raw brown sugar into a pot.

06 Continue to stir while bringing the sauce to boil.

07 Add cucumbers in to the boil sauce, turn to low heat, and then cook it for 10minutes.

08 Mix until cucumbers have colored and boiled.

09 Separate the sauce and cucumbers, until both sauce and cucumbers have cooled fully.

10 Than combine it together into a clean jar. Can refrigerate it up to a week.

Recipe 02

千層餅

Qian Cheng Savory Cake

備註

◆ 若覺得素肉燥較麻煩，可以不用加入。

◆ 若內餡想用甜的，也可改成甜食，或是調整為自己喜歡的內餡。

◆ 材料

紅蘿蔔	¼ 條	醬油	1～2 大匙
馬鈴薯	½ 顆	醬油膏	1～2 大匙
黑香菇	2～3 朵	黑胡椒	½ 大匙
高麗菜	⅕ 粒	糖	少許
素料	少許	素沙茶醬	1～2 小匙
素肉燥	½ 杯	麵粉	1～2 杯
油	2 大匙	水或高湯	1～1½ 杯

◆ 工具

擀麵棍
小刀子

◆ 做法

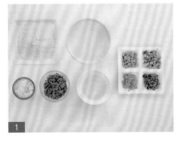

1　將所有的蔬菜洗淨、削皮、切至小碎丁，越小越好。然後用水沖洗素肉燥以及黑香菇後，浸泡至軟。濾乾後切成小碎丁。放旁備用。

2　取一個容器，加入麵粉、水、少許的糖、油，混合至均勻，呈麵團樣，用保鮮膜包起，放置旁備用。

3　熱鍋後轉小火加入油，當油熱後，加入黑香菇以及素肉燥拌炒成金黃色，然後將紅蘿蔔、馬鈴薯、高麗菜以及素料炒至食材均勻混合，加入醬油、醬油膏、黑胡椒、素沙茶醬。

4 加入水並煮滾後，將蔬菜煮至軟，並確認湯汁已收乾。關火，備用。

5 將麵團擀成長方形，約 0.3 公分厚，也可以盡量擀薄，愈薄的皮，在煎的時候，麵皮愈容易熟透。

6 將剛炒好的餡料用湯匙或是果醬刀，平均鋪在麵皮上。

7 用小刀在麵皮上、下 ⅓ 處切兩刀，分為三等分。

8 將最左下的餅皮摺至左邊中間。

9 再將左上的餅皮，摺至左側中間。

10 左邊的餅皮形成一個三層的正方形。最後再將左邊的餅皮往最中間摺。

11 中間的上層往下摺,摺至麵皮中心點。

12 然後中間的下層往上摺,摺至麵皮的中心點。

13 現在第二等分的麵皮應該已經形成一個六層的正方形。然後再將它往右邊摺至最後一等分的中間。

14 這時將最右邊的下層往上摺至中間。

15 最後將右邊的上層往下摺至中間。現在應該形成一個完整的正方形。這時候要準備煎。在煎之前,確定餅沒有太厚,如果太厚,要用手掌心將它壓扁一點。在煎的過程中,麵皮的中間夾層才會熟透。

16 在平底鍋加油後轉小火。

17 將剛包好的千層餅,放入中央,煎到雙面呈金黃色,取出後即可享用。

Qian Cheng Savory Cake

◆ INGREDIENT

Carrot	¼ whole
Potato	½ whole
Chinese black mushrooms	2-3 whole
Cabbage	⅓ whole
Soy bean product/meats	some
Textured vegetable protein (TVP)	½ cup
Oil	2T

Soy sauce	1-2T
Soy paste	1-2T
Black pepper	½T
Sugar	some
Vegetarian BBQ sauce	1-2t
Plain flour	1-2 cups
Water or vegetable stock/ broth	1-1 ½ cups

◆ TOOL

Rolling pin
Small knife

◆ METHOD

01　Wash, peel and cut all vegetables into fine cubes, as fine as you can cut. Wash and soak TVP and Chinese black mushrooms until soft, drain, cut into fine cubes. Set aside ready to use.

02　Pour plain flour into a bowl, add water, bit of sugar, and bit of oil, and mix well into dough, cover it with glade wrap, set aside to rest.

03　Heat the pan, turn to low heat, and add oil, when oil is hot, add Chinese black mushrooms and TVP in, stir fried it until golden, than add carrots, potatoes, cabbages and soy bean product/meats. Stir fried it till it is well mix. Add soy sauce, soy paste, black pepper, vegetarian BBQ sauce.

04　Add some water, let it bring to boil and cook till all ingredients are soft and the sauce/juice have all been absorbed. Then remove from stove, ready to use.

05　Roll out the dough into a rectangle shape around 0.3cm thickness, or a thin as you can, as this allow savory cake cook through.

06　Paste the filling on top of the dough sheet, evenly spread. Use a spoon or knife.

07　Cut two lines on the longer side of the rectangle dough (both sides), ONLY CUT ⅓ INTO THE DOUGH, evenly distributed apart between each cut.

08　From the left side, fold, the bottom into the middle.

09　Then follow by the top into the middle.

10　The first ⅓ of the dough should become a small square of three layers, than fold, the square towards the middle of the second ⅓ of the rectangle.

11　Then follow by the middle section top into the center of the dough sheet.

12　Then the middle section bottom into the center of the dough sheet.

13　The second ⅓ of the dough should become a small square of six layers, then fold, the square towards the last section of the last ⅓ of the rectangle sheet.

14　Then fold the bottom into the middle in the last section.

15　Last of all, fold the top into the middle in the last section. Than it is finished, ready to cook. Make sure it isn't too thick, if it is, use palm of the hand, and flatten it a bit more. So it can cook through the layers.

16　Heat a pan, add oil, and turn to very low heat.

17　Add the wrapped savory cake dough into the pan, fried till both side golden. Than take it out, ready to serve.

🍞 NOTE

• You don't need to use TPV if it is too much work.

• The filling can change to sweet or your own choice of filling.

◆ 材 料

麵粉	2 杯	油	少許
冷水	1 ～ 1½ 杯	鹽巴	少許
熱水	¼ ～ ½ 杯	香椿	少許

◆ 工 具

擀麵棍

香椿餅

Toona Chinese Pancake

◆ 做法

1　準備所有材料。

2　先將香椿洗、擦乾後,將香椿葉從樹枝或是梗上剝下。

3　將香椿葉剁碎。

4　將麵粉倒入一個容器中。

5　一邊攪拌一邊加入熱水,等到熱水與麵粉混合後,再加入冷水,直至呈團。

6　將攪拌均勻的麵團,倒出並放置到一個乾淨的平面。

7　用手掌將麵團搓揉至光滑。擦少許油至麵團表皮。

8　用保鮮膜包住醒 15 ～ 20 分鐘。

9　將麵團分為幾個手掌大後,用擀麵棍擀成 0.2 公分厚,也可以盡量擀薄,愈薄的皮,在煎的時候,麵皮愈容易熟透。

10 塗一層薄油至麵皮上，以及撒少許的鹽巴至麵皮上後，撒上少許的香椿末。

11 將麵皮捲起或往外摺，麵皮摺愈小，形成細長條狀，之後捲起的層次會愈多。

12 再將長條狀的麵皮往內捲成蝸牛殼狀的麵團，用保鮮膜包起來，醒 5 分鐘。

13 將食材拿出，用手掌心將麵團壓平，並擀成圓形或是方形。擀至 0.2 公分厚度。如果你喜歡軟一點不要很酥，可以不用擀的太薄。如果要酥一點的就可以擀薄一點。

14 熱鍋後加油，將餅皮放入鍋內，將兩面都煎至金黃色後，即可取出享用。

🍞 備註

◆ 若不喜歡香椿，也可做成原味，或是可改成你喜歡的香料做調味。

◆ 當使用越多熱水，麵團會愈軟，加入熱水主要是要讓麵團的筋性不要太強，以免口感過度有嚼勁，但若熱水也不宜加太多，以免變成糊狀。

◆ 當水加較多，麵團會越軟；若加較少，麵團會越硬。

Toona Chinese Pancake

◆ INGREDIENT

Plain flour	2 cups
Cold water	1-1 ½ cups
Hot water	¼- ½ cup
Oil	some
Salt	some
Toona	some

◆ TOOL

Rolling pin

◆ METHOD

01 Prepare all ingredients.

02 Wash, dry and peel the toona leaves from the branch/stem.

03 Dice the toona leaves into very fine pieces.

04 Add plain flour to a bowl.

05 Stir as you add hot water in. Mix it until hot water has been absorbed. Then add the cold water gradually, till it has becomes a dough.

06 Mix it well inside the bowl. When it is mixed, pour it onto a flat surface.

07 Use your palm and knead the dough until it is completely mix with smooth surface. Rub a thin layer of oil on the surface of the dough.

08 Cover it, and let it set for 15-20minutes.

09 Cut dough into palm size; roll each dough around 0.2cm thickness or as thin as you can.

10 Then spread a thin layer of oil on the surface, and sprinkle pinch of salt evenly on the surface after the oil, then a bit of the diced toona leaves, evenly on the very top.

11 Roll/fold it away from you, as small fold as you can. The smaller folds you make mean the more layers you will have. When it becomes a strip.

12 Then roll it sideways, into a snail bun. Cover. Let it rest for 5minutes.

13 Take out one of the snail dough; use hand of your palm, press it flat, roll it out into a flat circle/square. Around 0.2cm. If you don't like it too crispy, you don't need to roll it out thinly. But if you prefer crispy than roll it out as thin as you can.

14 Heat a pan, add oil, on low heat, and add the flat dough into the pan, fried till golden on both sides, plate up to serve.

☞ NOTE

- If you don't like toona, you can always have it plain, or add other flavoring that you may like.

- The more hot water you use, the softer the dough will become. Using hot water is to deactivate the gluten, so when you are making the dough, there will be less tension, less chewy to the final product, but you cannot use too much hot water otherwise, there will be just mush.

- The softer of the dough (mean the more water you add), the texture will be softer, the less water you add, the texture will be harder to chew, more texture when having it, which is how some people like it.

Recipe 04

蔬菜煎餅

Vegetable Chinese Pancake

◆ 材料

紅蘿蔔	⅕ 條	水	1½～2 杯
高麗菜	⅕ 粒	香菇味精	1～2 小匙
素料	⅓ 杯	鹽巴	⅓ 大匙
青椒	½ 粒	糖	1 小匙
洋菇	4 粒	白胡椒	少許
麵粉	1 杯	油	2～4 大匙

◆ 做法

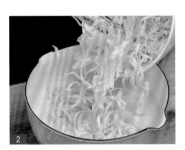

1　將所有的蔬菜洗淨、削皮、切成細絲狀，洋菇切片。素料切絲。

2　將高麗菜放入容器內。

3　加入紅蘿蔔。

4　加入素料。

5 加入青椒。

6 加入洋菇。

7 將麵粉與水混合均勻,並加入香菇
   味精、鹽巴、糖、白胡椒,混合均
   勻後,再倒入已混合的蔬菜,攪拌
   均勻後備用。

8 熱鍋後加油,待油熱後轉至小火。

9 加入剛混合好的食材,均勻的鋪平。

10 煎至邊緣呈現透明狀或是與麵糊的
   顏色不一樣,底部呈金黃色,再翻
   面,煎至雙面都呈金黃色。

11 起鍋、裝盤,表面可淋上自己喜歡
   的醬汁或調味料(醬油膏、甜辣醬、
   番茄醬、美乃滋、甜醬油等等),即
   可享用。

☺ 備註

◆ 麵糊需調成濃稠狀,不能太稀,否則在煎的時候食物不易成形。

◆ 麵糊與蔬菜的比例,應該為 2:1 或 1.5:1。

◆ 若你不想加入任何的醬,那你的麵糊需依個人喜好調味。

Vegetable Chinese Pancake

◆ INGREDIENT

Carrot	⅓ whole	Water	1 ½-2 cups
Cabbage	⅓ whole	Dried mushroom stock	1-2t
Soy bean product/meats	⅓ cup	Salt	⅓ T
Green capsicum	½ whole	Sugar	1t
Mushrooms	4 whole	White pepper	some
Plain flour	1 cup	Oil	2-4T

◆ METHOD

01 Wash, peel, and cut all vegetables and soy bean product/meats into thin shreds.

02 Add the cabbages into a container.

03 Add in carrots.

04 Add in soy bean product/meats.

05 Add in the green capsicums.

06 Add in the mushrooms.

07 Mix the plain flour and water into a pot, add dried mushroom stock, salt, sugar, white pepper into it, mix it well. Then pour it into the mix vegetables.

08 Heat a pan, add oil, when the oil is hot, and turn to low heat.

09 Pour the mixture into the pan; spread it evenly around the pan.

10 Fried till the edge has start to form and becomes a different color to the mixture and bottom has becomes golden, flip it over, and fried the other side the same.

11 When it's ready, plate up and ready to serve. Add your own desire dressing/sauce to go with it (soy paste, sweet chili, tomato sauce, mayonnaise, sweets soy paste etc).

⌂ NOTE

◆ Make sure the plain flour mixture is thick, not running, otherwise is going to be difficult to pull the vegetable together.

◆ The plain flour mixture needs to be more than the vegetables, ratio of 2:1 or 1.5:1.

◆ If you don't like to have any sauce, than make sure your flavoring in the plain flour mixture is enough.

Recipe 05

義大利麵沙拉

Pasta Salad

◆ 材料

迷你義大利麵	1 杯	油	1 大匙
馬鈴薯	2～4 個	美乃滋	1～2 杯
玉米粒	¼ 杯	鹽巴	1～1½ 大匙
青豌豆	¼ 杯	黑胡椒	少許
紅蘿蔔	¼ 杯	檸檬汁	依個人喜好
素火腿	½ 杯	巴西里	少許
水煮蛋	1～3 粒		

◆ 做法

1　將馬鈴薯與紅蘿蔔洗淨、削皮、切丁，水煮蛋與素火腿切丁，放旁備用。準備所有食材。

2　熱鍋後加油，轉中火，加入素火腿，炒至金黃色，放旁備用。

3　備一鍋水，煮滾後，倒入馬鈴薯並煮軟。

4　將馬鈴薯撈起，沖冷水並濾乾，放涼備用。將紅蘿蔔、玉米粒、青豌豆以相同程序煮 3～5 分鐘後，濾乾，冷卻放旁備用。

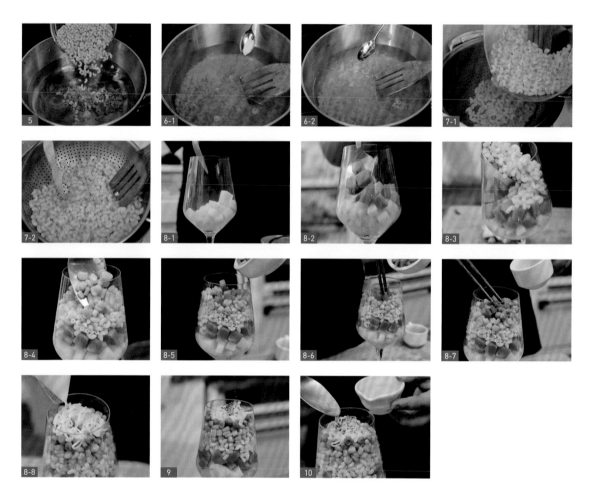

5 再煮另一鍋水，待滾後倒入迷你義大
　利麵。

6 加入油跟鹽巴，攪拌一下，滾 3 ～ 5
　分鐘，蓋鍋蓋、關火，悶約 5 分鐘，
　煮至迷你義大利麵變軟。

7 先將迷你義大利麵濾乾後，再次沖冷
　水及濾乾，並放旁備用。

8 取杯子或其他透明容器，依序加入馬
　鈴薯、紅蘿蔔、迷你義大利麵、玉米
　粒、素火腿、青豌豆、水煮蛋，在最
　上層擠上美乃滋。

9 灑上少許巴西里。

10 最後淋上檸檬汁，灑上鹽巴、黑胡椒。
　即可享用。也可以將馬鈴薯、紅蘿蔔、
　迷你義大利麵、玉米粒、素火腿、青
　碗豆、水煮蛋和巴西里、美乃滋、鹽
　巴、黑胡椒、檸檬汁放在容器內，攪
　拌並將食材混合均勻後享用。

Pasta Salad

♦ INGREDIENT

Small pasta	1 cup	Oil	1T
Potatoes	2-4 whole	Mayonnaise	1-2 cups
Frozen corns	¼ cup	Salt	1-1 ½T
Peas	¼ cup	Black pepper	some
Carrots	¼ cup	Lemon juice	as needed
Soybean-curd ham	½ cup	Parsley	some
Hard boil eggs	1-3 whole		

♦ METHOD

01 Wash, peel and cut the potatoes, carrots, hard boil eggs and soybean-curd ham into cubes. Set aside to use. Prepare all ingredients.

02 Heat a pan, add oil, medium heat, add the soybean-curd ham, stir fried till golden, set aside to cool, ready to use.

03 Bring a pot of water to boil, add in the potatoes, and cook till soft.

04 Drain, rinse under cold water, drain again, let it cool, ready to use. Then cook the carrots, frozen corns and peas with same process, for 3-5minutes, drain, cool and set aside ready to use.

05 Bring another pot of water to boil, add in the small pasta.

06 Add salt and oil into the pot, give it a stir, let it boil for 3-5minutes, cover it with lid, turn off the heat, let it soak around 5minutes, until small pasta are soft.

07 Drain, rinse under cool water. Drain, set aside ready to use.

08 In a container, add the potatoes, carrots, small pasta, frozen corns, soybean-curd ham, peas, hard boil eggs and finish off with mayonnaise.

09 Top it off with parsley.

10 Add the lemon juice, salt and black pepper as desire. Ready to serve. Or you can get a big container; add in the potatoes, carrots, small pasta, frozen corns, soybean-curd ham, peas, hard boil eggs and the sauce: parsley, mayonnaise, salt, black pepper, lemon juice. Combine all together. Mix it well. Then dish up ready to serve.

🍲 NOTE

• Hard boil eggs: you place the eggs into a pot, add cold water, cover it with lid, place on stove, cook till water is boiling (when the side of the lid has steam coming out), DO NOT OPEN THE LID, let it cook for 3minutes, turn off the heat let it soak for 5-7minutes. Then run under cold water, to peel of the shells. Cut the eggs into cubes of own desire. Set aside to use.

• Can add other vegetables.

Recipe 06

蔬菜沙拉

Cold Salad

◆ 材料

生菜	½ 個	蘋果	1 個
小黃瓜	¼ 個	紅蘿蔔	¼ 條
起司／菲達起司	少許	香菜	少許
番茄	2～3 個	鹽巴	1 大匙
玉米粒	½ 杯	黑胡椒	½ 大匙
青椒、紅椒、黃椒	½ 杯	橄欖油	3～4 大匙
水煮蛋（切片或塊）	1～3 個	義大利巴撒米克香醋	1～2 大匙

◆ 做法

1　依個人喜好將所有蔬菜、水果洗淨後，切塊或是切片、切條都可以。（註：切成同樣形狀，或是不同形狀會有不一樣的視覺效果。）

2　將橄欖油加入容器內。

3　加入義大利巴撒米克香醋，混合均勻後，放旁備用，待要用餐時再淋上食用。

4　準備一個可以上桌的漂亮容器，將生菜放在最下層，加入紅椒、黃椒、青椒，用顏色來襯托餐點的美感。

5　加入紅蘿蔔與小黃瓜。

6　加入蘋果片。

7　加入番茄片。

8　加入玉米粒以及水煮蛋。

9　加入起司與香菜。

10　食用前，加入鹽巴與黑胡椒後，在沙拉上淋醬汁，即可享用。如果太早淋上醬汁，生菜以及一些較會出水的菜會開始出水，而使沙拉變得軟爛，看起來感覺不新鮮。

🍞 備註

◆ 醬汁選擇：

　1.新鮮辣椒、鹽巴、胡椒、檸檬汁、醬油、橄欖油（可加入糖或甜醬油）。

　2.檸檬汁、鹽巴、胡椒、剁碎的薄荷葉（可依個人喜好加入新鮮辣椒）。

　3.蘋果醋、鹽巴、胡椒、花生或任何堅果類或是花生醬（可依個人喜好加入新鮮辣椒、醬油或芝麻香油）。

◆ 你可以加不同的食材在沙拉裡面，像是醃黃瓜、橄欖、薄荷、豆類（如果是罐頭，最好先煮一下）、堅果、水果、烤過的麵包丁，炸的食材像是炸蛋或是素料。

Cold Salad

♦ INGREDIENT

Lettuce	½ whole	Apple	1 whole
Cucumber	¼ whole	Carrot	¼ whole
Cheese/feta cheese	some	Coriander	some
Tomatoes	2-3 whole	Salt	1T
Frozen corns	½ cup	Black pepper	½T
Green, red and yellow capsicums	½ cup	Olive oil	3-4T
Hard boil eggs (slices or chunks)	1-3 whole	Balsamic vinegar	1-2T

♦ METHOD

01 Wash and cut all vegetables and fruit into chunks or slice or julienne, cut how you desired. (Sometime it's nice when it is same shape and size, sometime it's different with different cutting, give it more texture looks.)

02 Add olive oil in a bowl.

03 Then add in the balsamic vinegar. Mix well together. Set aside ready to use before serving.

04 Get a serving container, place lettuces on the bottom. Add all the capsicums in color sections. Use the ingredient colors to bring each individual ingredients stand out.

05 Add in the carrots and cucumbers.

06 Add in the apple slices.

07 Add in the tomato slices.

08 Add the frozen corns and hard boil eggs.

09 Finish with cheese and corianders.

10 Before serving, add salt and black pepper with the premade sauce into the salad evenly. If you add the sauces in to early, the lettuces and few other ingredients will start to release the vegetable liquids/juice, which will make the salad very soggy and soft, which will make the salad look not fresh. Ready to serve.

🍳 NOTE

♦ Sauce options:

1. Fresh chili, salt, pepper, lemon juice, soy sauce, olive oil (can use sugar or sweet soy sauce instead).

2. Lemon juice, salt, pepper, diced mint leaves (fresh chili).

3. Apple vinegar, salt, pepper, peanuts/nuts or even with peanut paste (fresh chili, soy sauce, sesame oil).

♦ You can add any type of ingredient or vegetables to your salads, can also add pickles, olives, mint, beans (if it's tin than you will need to be cook beforehand), nuts, fruits, toasted bread cubes or even deep fried ingredients example, deep fried eggs or crispy soy bean product/meats.

主
菜

Recipe 01

蔬菜義式烘蛋

Frittata

◆ 材料

白花椰菜	½ 杯	馬鈴薯	1 ～ 3 顆
綠花椰菜	½ 杯	馬蘇里拉起司	1 ～ 2 杯
節瓜或南瓜	½ 杯	雞蛋	6 ～ 8 顆
紅蘿蔔	¼ 杯	水	2 ～ 4 大匙
青豌豆	¼ 杯	鹽巴	少許
玉米粒	¼ 杯	胡椒	少許
番茄	1 ～ 1½ 粒	香菇味精	少許

◆ 工具

深底烤盤

◆ 做法

1　將白花椰菜、綠花椰菜、節瓜或南瓜、馬鈴薯切丁或切片，番茄需切片，約 0.5 公分厚，放置旁邊備用。

2　先將水煮滾後，將切丁的蔬菜（除了番茄以外）放入滾水中，煮至熟、軟。

3　將燙軟的蔬菜沖冷水後，瀝乾。

4 　將水、鹽巴、胡椒、香菇味精加入雞蛋中充分混合，並拌勻。

5 　將所有蔬菜充分混合，倒入深底烤盤中。

6 　倒入剛混合好的蛋液，均勻的倒入深底烤盤中與蔬菜混合。

7 　將馬蘇里拉起司均勻分布在食材表面，最後將切好的番茄片鋪在起司上。

8 　烤箱預熱至 180℃後放入烤盤，烤約 35 ～ 40 分鐘，烤成金黃色表面後，用
　　竹籤插入中心點，看有無沾黏，若沒有，就代表完全熟了，即可享用。

9 　如果要存放，要記得等待食材冷卻後，才可放置冰箱。

🍞 備註

◆ 可以用鮮奶油或鮮奶代替水，以增加濃郁感。

◆ 食材放置大烤盤中時，不可高過 2.5 公分，因為這樣中心點會不易熟。

◆ 不要加太多水，因為蔬菜裡面已有一些水分。

◆ 可以用烤馬芬的烤盤來代替一般烤盤，可便於放置冷凍庫存放約 3 個月。

Frittata

Cauliflower .... ½ cup

Broccoli .... ½ cup

Zucchini or pumpkin

.... ½ cup

Carrot .... ¼ cup

Peas .... ¼ cup

Frozen corns .... ¼ cup

Tomatoes .... 1-1 ½ whole

Potatoes .... 1-3 whole

Mozzarella cheese .... 1-2 cups

Eggs .... 6-8 whole eggs

Water 2-4T

Salt .... some

Pepper .... some

Dried mushroom stock .... some

◆ TOOL

Baking tray with a depth

◆ METHOD

01 Cauliflower, broccoli, zucchini or pumpkin, potatoes need to cut into chunks or slices. Tomatoes cut into 0.5cm slice, place aside to be use later.

02 Boil the vegetable till it's soft, except tomatoes.

03 Rinse in cold water and drain the vegetables.

04 Add water, dried mushroom stock, salt and pepper to the egg mixture, and mix it well.

05 Than mix the vegetable and place into the baking tray with a depth.

06 Pour in the egg mixture, level it.

07 Place mozzarella cheese on top, and finish it with tomato slices.

08 Place in oven at 180degree, for 35-40minutes. Till golden. Use a skewer to poke into the middle of the dish to check if the dish is cook thoroughly. The skewer should be clean after taken out from the dish.

09 If like to freeze the dish, wait till it cools and place in container, than place in the freezer.

♙ NOTE

• Can use thick cream or whip cream or milk instead of water, it will give more richness to the dish.

• Dish cannot be thicker than 2.5cm, if it's in a large tray, otherwise it won't be able to cook properly.

• Don't add to much water, as the vegetables has some water left.

• Can bake in muffin trays, for a better storage in the freezer, can store up to 3months.

Recipe 02

菠菜鹹派

Spinach Quiche

◆ 材料

A

鮮奶油	150 毫升
雞蛋	3 顆
鹽巴	2 小匙
黑胡椒	2 小匙
香菇味精	少許
中筋麵粉	3 大匙

B

油	2 大匙
菠菜	300 克
洋菇	4 大朵
玉米粒	50 克
素火腿	50 克
起司	2 ～ 3 杯

◆ 工具

深底烤盤
烤盤紙

◆ 做法

1　將所有蔬菜清洗乾淨後，把洋菇、素火腿切薄片厚度約 0.5 公分，素火腿請切與洋菇差不多大。

2　將水煮滾後，放入菠菜，燙約 3 ～ 5 分鐘後。

3　撈起。

4　沖冷水後濾乾後，放旁備用。

5　熱鍋後加入油，待油熱後加入素火腿，炒至素火腿邊緣呈金黃色。

6　加入洋菇。

7　加入菠菜。

8　加入玉米粒，均勻拌炒約5分鐘後關火。

9　將 A 部分所有材料混合後，並攪拌均勻，再將所有食材（A+B）充分混合。

10　加入起司至食材中。

11　將烤盤紙放入深底烤盤，並將食材倒入深底烤盤中，均勻鋪平。

12　可在最頂層依個人喜好加入起司。

13　烤箱預熱至 180℃後放入烤盤，烤約 20 分鐘，烤至金黃色後，用竹籤插入中心點，看有無沾黏，若沒有，就代表完全熟了，即可享用。

14　如果要存放，要記得等待食材冷卻後，才可放置冰箱。

Spinach Quiche

◆ INGREDIENT

A

Thick cream or whip cream	150ml
Eggs	3 whole eggs
Salt	2t
Black pepper	2t
Dried mushroom stock	some
Plain flour	3T

B

Oil	2T
Spinach	300g
Mushrooms	4 whole
Frozen corns	50g
Soybean-curd ham	50g
Cheese	2-3 cups

◆ TOOL

Baking tray with a depth

Baking paper

◆ METHOD

01 Wash all vegetables, cut mushrooms and soybean-curd ham into strips of 0.5cm thick. Soybean-curd ham should be around same size as the mushrooms.

02 Bring water to boil, add spinach in and cook for 3-5minutes.

03 Drain.

04 Raised under cold water, drain and place aside ready to use.

05 Heat the pan, add the oil, when it's heated, add the soybean-curd ham. And cook till golden on the edge.

06 Add the mushrooms.

07 Add spinach.

08 Add frozen corns, cook for 5minutes, than turn off the heat.

09 Add all ingredient A mixture together, mix well. Then pour the mixture into ingredient B. Mix well.

10 Add cheese into mixture.

11 Place baking paper on the baking tray with a depth, then pour the mixture into it, level evenly.

12 Add more cheese to the top to finish.

13 Place into oven on 180degree, for 20mintues, till golden. Use a skewer to poke into the middle of the dish to check if the dish is cook thoroughly. The skewer should be clean after taken out from the dish. Ready to serve.

14 If like to freeze the dish, wait till it cools and place in container, than place in the freezer.

Recipe 03

蔬菜餅／堡

Veggie Patty Burger

◆ 材料

馬鈴薯	4 粒	雞蛋	2～4 粒
三色豆（冷凍青豌豆、玉米、紅蘿蔔）	½～1 杯	奶油	1½～3 大匙
		鹽巴	½ 大匙
麵包屑	1½～3 杯	胡椒	少許

◆ 工具

湯匙或叉子

◆ 做法

1 馬鈴薯削皮切薄片，放置滾水中，燙至軟，撈起後濾乾。

2 將水煮滾，放入三色豆燙 3 分鐘後，撈起濾乾，放置旁邊備用。

3 將馬鈴薯片用湯匙或叉子壓成泥。

4 將奶油加入馬鈴薯混合均勻。

5 待馬鈴薯未涼，加入鹽巴、胡椒，攪拌均勻。

6 混合馬鈴薯及三色豆，並攪拌均勻。

7 加入麵包屑，攪拌均勻。

8 在碗中將雞蛋攪拌均勻，並拿一個碗放入麵包屑，放旁備用。

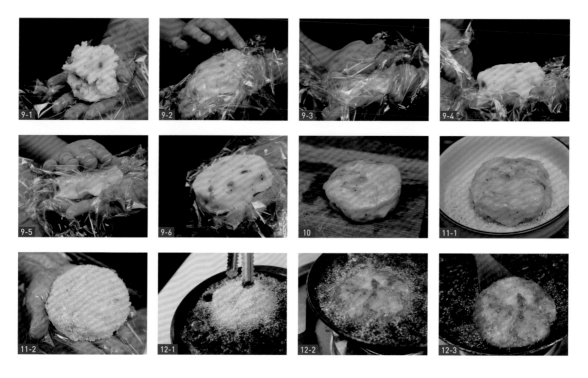

9 將食材取一些約手掌大小，滾至圓球後壓扁並緊實，像漢堡肉形狀（約 1 公分厚）。

10 沾取蛋液。

11 再均勻的沾取麵包屑。

12 熱油後沿鍋邊放入，炸或煎至兩面都呈金黃色後取出。

13 可直接包入漢堡使用，或放涼備用。

✿ 備註

◆ 馬鈴薯要趁熱壓成泥，若冷卻後，會較難處理，不容易弄均勻。

◆ 可以放入冷凍庫備用，第一次煎不要煎的太金黃色，因為使用前會再煎一次，可存放 3 個月。

◆ 麵包屑是馬鈴薯泥跟蔬菜混合在一起時可協助成形，所以加入時，要慢慢的加入比較好控制，如果加的不夠多，比較不容易成形也不好操作。可是如果加太多，會變成太乾，容易裂開。

Veggie Patty Burger

◆ INGREDIENT

Potatoes	4 whole	Eggs	2-4 whole eggs
Mix of frozen pea, corn and carrot	½-1 cup	Butter	1 ½-3T
Bread crumbles	1 ½-3 cups	Salt	½T
		Pepper	some

◆ TOOL

Spoon or fork

◆ METHOD

01 Peel potatoes, cut into thin slices, boil potatoes till very soft, drain.

02 Boil the mix of frozen pea, corn and carrot for 3minutes, drain and place aside for use.

03 Mash the potatoes with a masher or a spoon/fork.

04 Add butter to the mash potatoes, mix well.

05 Add salt and pepper, mix well while it is still hot.

06 Add mix of frozen pea, corn and carrot into the mash potato mixture.

07 Add bread crumbles, mix well.

08 Beat some eggs in a bowl, and place some bread crumbles in the other bowl, set aside to be use.

09 Scoop palm size filling, roll into a round tight ball shape, and flat it into a patty, 1cm thick, round disk.

10 Dip it into egg mixture.

11 Then dip into the bread crumbles evenly.

12 Heat the oil, place it into the hot oil, deep fried it or pan fry till golden on both side.

13 Can serve in a burger or set aside to cool.

🍲 NOTE

• The potatoes need to be mash while is still hot, because when it's cold, it will be hard to mash and mix well.

• If like to place in freezer for later use, make sure the first time don't fried the patty too golden, so the time when you like to serve, can you warm up till golden to serve while warm. It can be keep in the freezer up to 3months.

• Bread crumble is to help the mixture stay dry and stick together, add small amount as you go, until it is easy to form its shape. If you add to much it will be to dry, and may fall apart as well.

◆ 材料

墨西哥餅皮	3 ～ 6 片
番茄	½ 顆
小黃瓜	⅓ 根
生菜	¼ 顆
洋菇	2 ～ 4 朵
茄子	½ 條

雞蛋	3 ～ 6 顆
素火腿	依個人喜好
鹽巴	少許
胡椒	少許
美乃滋	少許
油	1 ～ 2 大匙

◆ 工具

三明治製造機

Recipe 04

蔬菜墨西哥捲

Mexican Tortilla

◆ 做法

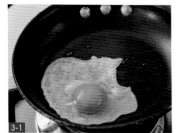

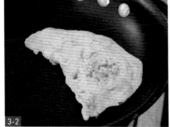

1　清洗所有蔬菜，將番茄、洋菇、小黃瓜、茄子、素火腿切薄片；將生菜切細絲。

2　熱鍋後，倒入油，加入素火腿、茄子、洋菇，分別將雙面炒至金黃色，撈起後放旁備用。

3　煎雞蛋（熟度依個人喜好），煎好放旁備用。

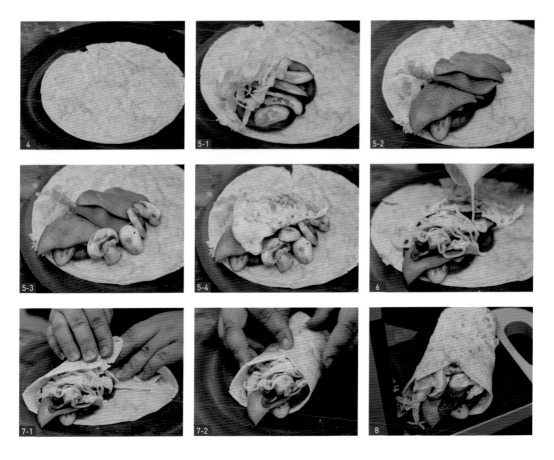

4　將墨西哥餅皮攤開，放置乾淨平面上。

5　將不同食材放置墨西哥餅皮中間。

6　加入美乃滋、鹽巴、胡椒，並依個人喜好加入其他調味料。

7　將墨西哥餅皮捲起：先將下面摺起並包覆食材後，再分別包左、右邊。

8　放入三明治製造機壓一下，待至燈轉換為綠色，或食材形狀固定即可享用。

🍳 備註

◆ 墨西哥餅皮的包法，可依個人喜好稍做改變。

◆ 放置中間的食材，可依個人喜好加入起司、紅蘿蔔等，但若購買的墨西哥餅皮較小，請不要放太多，否則不易包起，且會有裂開的狀況發生。

◆ 可以加入番茄醬、檸檬汁、新鮮辣椒等調味料，以增添風味。

Mexican Tortilla

◆ INGREDIENT

Mexican tortillas	3-6 piece	Eggs	3-6 whole eggs
Tomato	½ whole	Soybean-curd ham	as needed
Cucumber	⅓ whole	Salt	some
Lettuce	¼ whole	Pepper	some
Mushrooms	2-4 whole	Mayonnaise	some
Eggplant	½ whole	Oil	1-2T

◆ TOOL

Sandwich presser

◆ METHOD

01 Wash all vegetables, tomato, mushrooms, cucumber, eggplant and soybean-curd ham cut into thin slice. Cut lettuce into thin shreds.

02 Heat pan, add oil, than fried soybean-curd ham, eggplants and mushrooms separately till golden on both sides, set aside to be use.

03 Than fried the eggs till however you like. Set aside to be use.

04 Place the Mexican tortilla on a clean surface.

05 Add the ingredient into the center of the Mexican tortilla.

06 Add mayonnaise, salt, pepper, and other flavoring you like on top.

07 Wrap it up by folding the bottom first towards the center of the Mexican tortilla, follow by two either side.

08 Than place in the sandwich presser, and press, until light change, or it has be mold into shape. Serve hot.

🍳 NOTE

- When trying to wrap it up, you can wrap however you like, as long as the filling is not falling out.
- The center ingredient can be place or added cheese, carrot as you desire, but make sure if you have a small Mexican tortilla, don't add to much ingredient otherwise it will be difficult to wrap up, or it may break from the bottom.
- Can add tomato sauce or lemon juice or fresh chili to give more flavors.

Recipe 05

酥皮菜捲

Vegetarian Sausage Roll

◆ 材 料

酥皮	3～6 片	馬鈴薯	1 粒
豆腐	2～3 塊	素料（示範為素火腿）	少許
雞蛋	2～4 顆	起司	1 杯
三色豆（冷凍青豌豆、玉米、紅蘿蔔）	½ 杯	鹽巴	1 大匙
		胡椒	½ 大匙

◆ 工具

烤盤
烤盤紙
叉子

◆ 做法

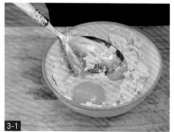

1 將馬鈴薯清洗後並削皮、切小丁；將素火腿切成小丁，將其他材料備齊。

2 把豆腐放入碗裡充分搗碎。

3 加入雞蛋，混合均勻。拌均勻後，加入胡椒、鹽巴調味。

4 　加入素火腿以及三色豆，混合均勻。

5 　加入馬鈴薯混合均勻。

6 　加入起司與所有材料混合均勻。

7 　將酥皮放置一個乾淨的桌面上，把
　　靠自己的那一邊，放入剛混合的食
　　材後，將食材捲起向外推。

8 　將叉子插入上層餅皮至食材中，但
　　需注意不可插穿至下層餅皮。

9 　捲好後，將成品放在已放烤盤紙的
　　烤盤上後，烤箱預熱至 200℃後放
　　入烤盤，烤約 15 ～ 20 分鐘，烤至
　　餅皮呈金黃色，即可取出享用。

🍮 備註

◆ 食譜會因為豆腐與馬鈴薯大小不同需
　要做少許的調整。

◆ 餡料不可太濕，若太濕會導致酥皮無
　法膨起。

◆ 若真的太濕，可倒掉一些液體，只要
　確保每個食材都有沾到起司跟蛋液，
　就不會裂開。

◆ 在包餡料的時候，注意動作要快，否
　則蛋液易流出。

◆ 也可以用扁豆以及三色豆做為內餡，
　首先先將扁豆煮軟，瀝乾後搗成泥，
　然後加入三色豆、少許麵包屑，混合
　均勻，包入酥皮內做為內餡。或是可
　用馬鈴薯代替扁豆；並按照食譜的包
　法以及烤的程序製作。

Vegetarian Sausage Roll

◆ INGREDIENT

Puff pastries	3-6 piece	
Tofu	2-3 piece	
Eggs	2-4 whole eggs	
Mix of frozen pea, corn and carrot	½ cup	
Potato	1 whole	

Soy bean product/meats	some
Cheese	1 cup
Salt	1T
Pepper	½T

◆ TOOL

Baking tray
Baking papper
Fork

◆ METHOD

01 Wash the potato peel and cut into small cubes. Cut soy bean product/meats into small cubes. Prepare all ingredients.

02 Place the tofu in a bowl, and mash it well.

03 Then add egg, salt and pepper, and evenly mix the ingredient.

04 Add soy bean product/meats and mix of frozen pea, corn and carrot into mixture.

05 Add potatoes and mix it well.

06 Add cheese into mixture, mix it well.

07 Place the puff pastry on a clean surface, place the mixture on one side of the puff pastry edge that is closer to you, and then roll it away from you with filling inside.

08 Use a fork to pierce through the top of the pasty roll, make sure it doesn't pierce through the bottom puff pastry.

09 When roll is finish, place the roll onto the baking paper, which is on the baking tray. Place into oven at 200degree, for 15-20minutes. Until golden. Ready to serve.

☺ NOTE

- Alternation may require due to different sizes of the tofu and potato.
- The mixture can't to be wet; otherwise the puff pastry will not puff.
- If it is too wet, than you can drain some liquid out, as long as all ingredient has a layer of eggs, and cheese is well mix, than it will hold it in place.
- You will need to wrap it up quickly; otherwise the eggs will leak out.
- You can also use lentil and vegetables ingredients as filling, first you cook the lentil till soft, drain, mash it then add in the mix of frozen pea, corn and carrot, with some bread crumbles, mix it well, then fill it into the puff pastry to become the fillings, or you can use mash potato instead of the lentils. Can follow the same way of rolling the puff pastry and bake.

Recipe 06

牧羊蔬菜派

Vegetarian Cottage Pie

◆ 材料

馬鈴薯	5～8 顆
三色豆（冷凍青豌豆、玉米、紅蘿蔔）	½～1 杯
洋菇	4 朵
番茄	4～5 粒
番茄泥	½ 杯
油漬番茄乾	6 粒
番茄汁焗豆	½～1 罐
大紅豆	½～1 罐
鷹嘴豆	½～1 罐

起司	1～3 杯
油	3 大匙
奶油	1～2 大匙
水	½ 杯
鹽巴	少許
胡椒	少許
香菇醬油膏	2～3 大匙
香菇味精	½ 大匙
義大利香料	少許
鮮奶	1～2 杯

◆ 工具

焗烤的碗或
深底烤盤

◆ 做法

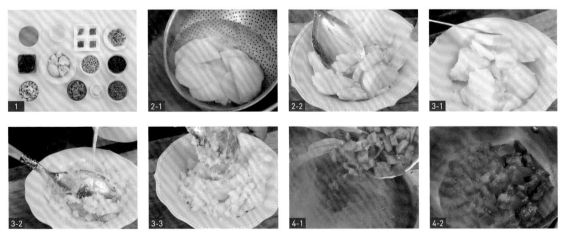

1　將所有食材清洗乾淨、切丁，並備齊所有材料。

2　煮水並加入馬鈴薯片，煮至軟、瀝乾，並均勻搗碎。

3　加入奶油、鮮奶混合均勻，放旁備用。

4　熱鍋後倒入油，待油熱後，加入番茄，炒至沸騰。

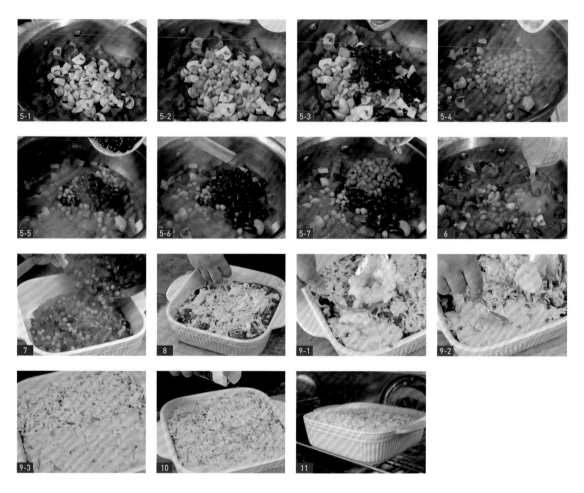

5 　加入洋菇以及三色豆、油漬番茄乾、鷹
　　嘴豆、大紅豆、番茄汁焗豆。

6 　加入番茄泥、鹽巴、胡椒、香菇醬油
　　膏、香菇味精、義大利香料，煮滾後轉
　　小火，熬煮至濃稠。

7 　將食材倒入焗烤的碗或深底烤盤中，約
　　七分滿。

8 　加入起司。

9 　放入馬鈴薯泥於頂端，並蓋過所有食材。

10 撒上少許胡椒。

11 烤箱預熱至 200℃後放入烤盤，烤約
　　10 ～ 20 分鐘，或烤至馬鈴薯呈金黃色，
　　即可取出享用。

🍳 備註

◆ 若想加入起司，可放置馬鈴薯裡，混合均勻，或放置在馬鈴薯與食材中的中間夾層處。

◆ 食材可依個人喜好改變。

Vegetarian Cottage Pie

◆ INGREDIENT

Potatoes	5-8 whole	Cheese	1-3 cups
Mix of frozen pea, corn and carrot	½-1 cup	Oil	3T
Mushrooms	4 whole	Butter	1-2T
Tomatoes	4-5 whole	Water	½ cup
Tomato puree	½ cup	Salt	some
Sundried tomatoes	6 whole	Pepper	some
Bake beans	½-1 can	Mushroom soy paste	2-3T
Red beans	½-1 can	Dried mushroom stock	½T
Chickpeas	½-1 can	Italian herbs	some
		Milk	1-2 cups

◆ TOOL

Baking bowl/tray with a depth

◆ METHOD

01 Wash all vegetables, and cut it into cubes. Prepare all ingredients.

02 Boil water, add potato slices, until soft, drain. Mash the potatoes.

03 Add butter and milk, mash it together till it becomes well mixed. Rest aside to be use.

04 Heat pot, add oil, when it is hot, and add tomatoes, fried until nearly boil.

05 Add mushrooms, the mix of frozen pea, corn and carrot, sundried tomatoes, chickpeas, red beans and bake beans.

06 Then add tomato puree, salt, pepper, mushroom soy paste, dried mushroom stock, Italian herbs, bring to boil. Turn it to low heat, and cook till it become thick.

07 Pour ingredient into a baking tray with a depth bowl. Fill it up half full or ⁷⁄₁₀.

08 Then add cheese.

09 Add the potato mash on top, cover the mixture.

10 Finish with pepper on top.

11 Place into oven at 200degree; bake 10-20minutes, till potatoes is slight golden. Serve hot.

 NOTE

◆ If like to have some cheese, it can be mix with the potato mash or cheese can be place under the mash, above the mixture.

◆ The ingredients can be change as own desire.

◆ **材 料**

中筋麵粉	2 杯	優格	1 ～ 1½ 杯

以下食材，可依個人喜好加入：

番茄	1 粒	墨西哥辣椒	½ 大匙
黃椒、青椒	¼ 杯	萬用番茄醬	底醬
洋菇片	2 朵	起司	½ ～ 1 杯
鳳梨	1 大匙	胡椒	少許
玉米粒	1 大匙	辣椒粉	少許

◆ **工 具**

擀麵棍

叉子

烤盤

烤盤紙

Recipe 07

披薩

Pizza

◆ 做法

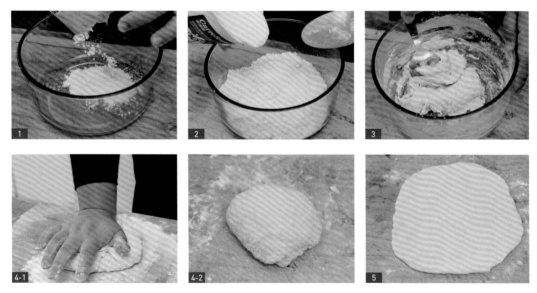

1　將中筋麵粉加入碗中。

2　分次將優格加入中筋麵粉中並拌勻，直至充分混合。

3　直到中筋麵粉成麵團。

4　將麵團放置一個平面上，用手揉至混合均勻。

5　用擀麵棍擀開麵團，厚度約 0.8 ～ 1 公分。

6 將擀開的麵團放在已放烤盤紙的烤盤上
   後，用叉子均勻的將麵團表面戳洞。

7 烤箱預熱至 200℃ 後放入烤盤，烤約 10
   分鐘，烤至餅皮邊緣處有些許金黃色即
   可取出，放涼備用。

8 將萬用番茄醬均勻抹在餅皮上，加入
   玉米粒、鳳梨、青椒、黃椒、番茄、
   墨西哥辣椒。

9 放入起司及胡椒。

10 加入辣椒粉。

11 最後放上洋菇片。

12 烤箱預熱至 200℃ 後放入烤盤，烤
   約 10 ～ 15 分鐘，烤至金黃色，即
   可取出享用。

🍳 備註

蔬菜不可放太厚，不可超過 2 公分，因為會不易熟。

Pizza

◆ INGREDIENT

Plain flour	2 cups
Greek yoghurt/plain yoghurt	1-1 ½ cups

Below are the ingredient you like to add for your pizza:

Tomato	1 whole	Jalapeno chili	½T
Yellow and green capsicum	¼ cup	Universal tomato sauce	as base sauce
Mushrooms	2 whole	Cheese	½ -1 cup
Pineapple	1T	Pepper	some
Frozen corns	1T	Chili flakes	some

◆ TOOL

Rolling pin

Fork

Baking tray

Baking paper

◆ METHOD

01 Add plain flour into a bowl.

02 Add greek yoghurt/plain yoghurt in small portion into the plain flour, add as you mix it with plain flour.

03 Until the mix becomes a dough.

04 Place the dough onto a flat surface, knead till it is well mix.

05 Use the rolling pin to roll out the dough into 0.8-1cm thick base.

06 Place the dough sheet on baking paper, which is on the baking tray. Use fork to evenly pierce on the top of the dough base.

07 Place in oven on 200degree for 10minutes till slight golden on the edge. Take out to cool for use.

08 Evenly spread universal tomato sauce on the pizza base, then add frozn corns, pineapples, yellow and green capsicums, tomatoes, jalapeno chili.

09 Place cheese and pepper on top.

10 Can add chilli flakes on the top.

11 Finish off with mushrooms on top.

12 Place in the oven at 200degree for 10-15minutes, until golden. Ready to serve.

🎩 NOTE

The topping can't be place to thick, no more than 2cm, otherwise the center of the pizza won't be cook thoroughly.

焗烤大白菜

Bake Chinese Cabbage

◆ 材料

油	3 大匙	胡椒	1½ 大匙
大白菜	1 粒	香菇味精	少許
紅蘿蔔	1½ 條	鮮奶油	250 克
洋菇	5 粒	酥皮	1 ～ 3 張
素料（或素火腿）	少許	雞蛋	5 ～ 7 顆
鹽巴	1 ～ 2 大匙		

◆ 工具

深底烤盤
烤盤紙

◆ 做法

1　清洗所有蔬菜，將大白菜、紅蘿蔔、素料切細絲，洋菇切片。

2　將雞蛋打入碗中，加入鮮奶油。

3　將雞蛋與鮮奶油混合均勻。

4　熱鍋後加入油。

5　加入紅蘿蔔。

6　加入大白菜。

7 加入洋菇。

8 加入素料。

9 將所有食材翻炒約五分鐘後,加入
   鹽巴、胡椒混合均勻。

10 加入香菇味精。

11 當所有蔬菜炒軟後關火,並加入已
   混合好的雞蛋與鮮奶油,與食材混合
   均勻。

12 將深底烤盤放入烤盤紙後,將食材
   倒入深底烤盤中。

13 把酥皮鋪在食材頂端,並完全覆蓋
   食材。

14 烤箱預熱至 160℃ ∼ 180℃ 後放入烤
   盤,烤約 40 ∼ 45 分鐘,烤至酥皮呈
   金黃色,即可取出享用。

🍞 備註

◆ 如果你想要焗烤有點嚼勁,可多加幾顆雞蛋,若你想要口感較軟,用 5 顆雞
  蛋即可。

◆ 炒菜後會有較多的湯汁,可將少許湯汁加入焗烤內,如果湯汁剩太多,不可
  全部加入,不然會無法成形。

Bake Chinese Cabbage

◆ INGREDIENT

Oil 3T

Chinese cabbage .... 1 whole

Carrots .... 1 ½ whole

Mushrooms .... 5 whole

Soy bean product/meats
.... some

Salt 1-2T

Pepper .... 1 ½T

Dried mushroom stock
.... some

Thick cream or whip cream
.... 250g

Puff pastries .... 1-3 sheets

Eggs .... 5-7 whole eggs

◆ TOOL

Baking tray with a depth

Baking paper

◆ METHOD

01 Wash all vegetables, cut Chinese cabbage, carrots and soy bean product/meats into thin shreds, mushrooms in slices.

02 Add eggs into a bowl, and pour thick cream or whip cream into the eggs.

03 Beat the eggs, and the thick cream or whip cream together, mix it well.

04 Heat the pot, medium heat, than add oil.

05 Add carrots into the pot.

06 Add Chinese cabbages.

07 Add mushrooms.

08 Add soy bean product/meats.

09 Stir fried all vegetables for 5minutes, than add salt and pepper, continue to stir fried.

10 Add dried mushroom stock.

11 Turn off the heat when all vegetable has become soft and add the egg cream mixture. Combine it together.

12 Place baking paper inside a baking tray with a depth, then pour the mixture into it.

13 Add the puff pastry on top, cover the mixture.

14 Place into oven on 160-180degree and bake for 40-45minutes till the puff pastry is golden and serve hot.

🍳 NOTE

♦ If you like it more solid, than add more eggs, but if like it more soft, than keep at 5eggs.

♦ There will be lots of liquid sauce after the stir fried vegetables, you can add some into mixture, but if there are a lot of liquid, make sure you don't add it all in, otherwise it will not form.

西式主菜 WESTERN MAIN COURSE

Recipe 09

柑橘白醬
義大利麵

White Pasta with Orange Tint

◆ 材 料

綠花椰菜	½ 杯	胡椒	½ 大匙
紅蘿蔔	⅓ 杯	起司	依個人喜好
洋菇	2 ～ 3 朵	柳橙絲或柳橙末	少許
香芹／荷蘭芹／巴西里	少許	義大利麵	½ ～ 1 包
鮮奶油	1½ ～ 3 杯	油	少許
鹽巴	1 ～ 1½ 大匙	奶油	1 大匙

◆ 做法

1 將所有食材，清洗、削皮；將洋菇、綠花椰菜、紅蘿蔔切成大塊或丁，備齊所有材料。

2 把水煮滾，加入蔬菜，待蔬菜熟後，撈起，加入冷水加以冷卻，放旁備用。

3 加入油與鹽巴至滾水中。

4 加入義大利麵，待水滾後，再繼續煮 3 分鐘，關火並蓋上蓋子，悶 5 分鐘，瀝乾，加入少許的油，拌均勻後放旁備用（油主要是要幫助義大利麵不要黏在一起）。

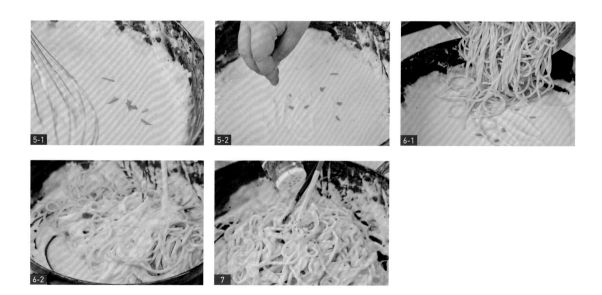

5 小火熱鍋後，加入奶油、鮮奶油、起
   司、胡椒、鹽巴，煮至濃稠後，加入
   柳橙絲或末。

6 將煮好的義大利麵加入，混合均勻。

7 加入少許的巴西里末。

8 最後加入剩下的柳橙絲或末，以及巴
   西里末在義大利麵的最上層。再將煮
   好的蔬菜放到義大利麵的旁邊，擺盤
   後即可享用。

🍴 備註

◆ 可使用香菇味精置於鮮奶油內一起烹煮，會更加美味。

◆ 義大利麵條大小的不同，會影響烹飪時間的長短，請依實際情況做調整。

◆ 煮義大利麵時，義大利麵與水的比例為 1：3。

◆ 如果你想讓食物的味道層次更豐富，則可以在白醬裡加入南瓜泥、馬鈴薯泥、菠
  菜泥或是青醬。首先你要煮一鍋水，加入你喜歡的蔬菜，當蔬菜變軟，就可以瀝
  乾後搗成泥，即可加入至白醬中混合均勻。但因青醬已經是做好的醬，可直接加
  入，不需要做其他動作。

White Pasta with Orange Tint

◆ INGREDIENT

Broccoli	½ cup	Pepper	½T
Carrot	⅓ cup	Cheese	as needed
Mushrooms	2-3whole	Orange grated rind	some
Parsley	some	Pasta noodles	½-1 pack
Thick cream or whip cream	1 ½-3 cups	Oil	some
Salt	1-1 ½T	Butter	1T

◆ METHOD

01 Wash and peel all vegetables and cut mushrooms, broccoli, carrot into chunks. Prepare all ingredients.

02 Bring a pot of water to boil, add in the vegetables cook until soft, drain, and risen under cool water, set aside ready to use.

03 Then add oil and salt into the pot of boiled water.

04 Add in the pasta noodles in, bring to boil than cook it for 3minutes, cover it pot with lid. Then turn off the heat, let it soak for 5minutes. Drain, add a bit of oil, and stir it in. Set aside ready to use (oil is to help the pasta not to stick together).

05 Use low heat, heat the pan, add in the butter and pour the thick cream or whip cream in, add in the cheese, pepper, salt, when it has thicken add orange grated rind.

06 Add the cook pasta noodles into the sauce. Mix well, plate up.

07 Add some parsley.

08 Add orange grated rind and rest of the parsley on top of the pasta noodles. Add the soft vegetables on the side, ready to serve.

🍴 NOTE

◆ Can add dried mushroom stock into the thick cream or whip cream while cooking. Can add more flavor to the dish.

◆ Size of the pasta noodles will also depend on the time to cook.

◆ While cooking pasta noodles, the pasta noodles and water ratio is 1:3.

◆ If you like to have a flavored sauce, you can add mashed pumpkin, or mash potato, or mash spinach, or you can add some pesto sauce. So you cook the vegetable you would like in boil water until soft, than drain it well, mash it smooth, than add it to the cream mixture. You won't need to do that with pesto sauce, as it is already ready to add into sauce.

♦ **材料**

馬鈴薯	4～6個
三色豆（冷凍青豌豆、玉米、紅蘿蔔）	½杯
奶油	2～3大匙

鮮奶	½杯
起司	1～2杯
鹽巴	1～2大匙
胡椒	1大匙

♦ **工具**

錫箔紙

烤盤紙

烤盤

叉子

Recipe 10

焗烤馬鈴薯

Bake Cheese Potato

◆ 做法

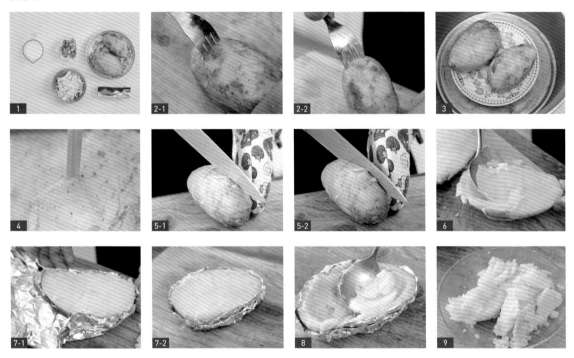

1　將馬鈴薯洗乾淨，準備所有材料。

2　用叉子將馬鈴薯的周圍刺一些洞。要刺到中心可是不要穿過去。這樣在蒸的時候馬鈴薯比較容易熟。

3　將刺好的馬鈴薯放在容器內，放入煮飯鍋中，蒸熟至馬鈴薯軟。如果你沒有煮飯鍋，你可以放在微波爐裡面蒸熟或是用鍋子蒸熟。

4　蒸至筷子可刺進馬鈴薯中心後，即確定裡面有熟。

5　用刀子將馬鈴薯對切。

6　用湯匙將馬鈴薯中心的馬鈴薯挖出，但不要挖破它的皮。

7　如果有挖破或是馬鈴薯沒辦法呈現碗形，你可以用錫箔紙包住馬鈴薯，讓它成形。

8　包住後，如果你覺得馬鈴薯殼太厚，可以再挖一些內餡出來，但以不影響外觀為主。

9　將挖出來的馬鈴薯放入一個容器內。

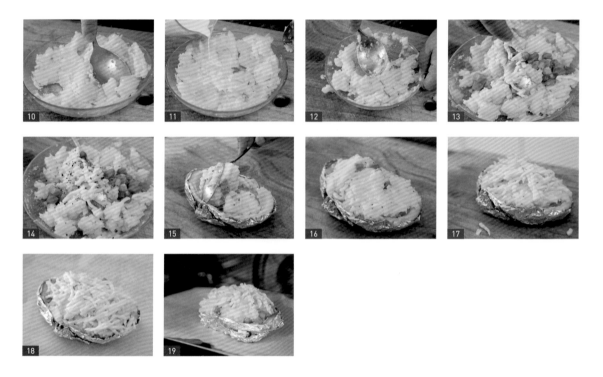

10 用湯匙或是叉子將馬鈴薯搗成泥。

11 加入奶油與鮮奶。

12 將鮮奶與馬鈴薯泥混合均勻成一個滑順的泥。

13 加入三色豆和起司。

14 加入鹽巴、胡椒，將所有食材攪拌均勻。

15 將拌好的馬鈴薯內餡舀進步驟 8 馬鈴薯外殼內。

16 裝滿馬鈴薯內餡，越多越好。

17 最後在上面撒上少許的起司。（註：可依個人喜好加入一些辛香料。）

18 先在烤盤上鋪烤盤紙，並將做好的馬鈴薯放在烤盤上。

19 烤箱預熱至 200℃ 後放入烤盤，烤約 10 ～ 15 分鐘，或是烤至起司融化，上面呈現金黃色為止。

20 取出後即可享用。

備註

◆ 如果不小心將馬鈴薯蒸過軟，馬鈴薯的皮會比較脆弱，但也有一些馬鈴薯的皮比較薄，一樣容易破，所以最好每 15 分鐘檢查一下。

◆ 你可以加入一些辣的調味料或是新鮮辣椒。

◆ 你可以依喜好改變內餡。

◆ 你可以用番茄代替馬鈴薯，若不希望番茄中間太多番茄汁，可以將中心挖出後，再加入自己喜歡的食材。

Bake Cheese Potato

◆ INGREDIENT

Potatoes	4-6 whole	Milk	½ cup	
Mix of frozen pea, corn and carrot	½ cup	Cheese	1-2 cups	
Butter	2-3T	Salt	1-2T	
		Pepper	1T	

◆ TOOL

Aluminum foil

Baking paper

Baking tray

Fork

◆ METHOD

01 Wash the potatoes nice and clean. Prepare all ingredients.

02 Use a fork, pierce through the potatoes all the way to the center of the potatoes, and pierce all around the potatoes, so it can cook through easily, but don't pierce through the potatoes.

03 Place the pierced potatoes on a plate and place it in a rice cooker or a steam, let it cook till it is soft. If you don't have a rice cooker, you can use microwave or steam it in a pot.

04 When it is cooked, use a chopstick; pierce through to check the center has been cooked properly.

05 Use a knife and cut it in half.

06 Use a spoon and scoop out the center of the potatoes. Try to scoop out as much as possible without breaking the potato skin.

07 If it does break or it's not holding, and then use aluminum foil, wrap around the potatoes, to hold potato's shape.

08 When the aluminum foil has hold the potato's shape, and you feel comfortable, you can scoop out more of the potato fillings.

09 Place the potatoes that has been scooped out into a bowl.

10 Use a spoon or fork to mash the potatoes.

11 Add milk and butter to the potato mash.

12 Mix it well, until it becomes a nice smooth paste.

13 Add the mix of frozen pea, corn and carrot, and cheese into the potato mixture.

14 Add salt, pepper. Combine everything together.

15 Scoop the fillings back inside the potato skin.

16 Fill it up as much as you can.

17 Finish the top with cheese. (Can add more pepper or herbs if desire.)

18 Place it on a baking tray with baking paper on the bottom.

19 Put it in the oven at 200degree for 10-15minutes, until the cheese has melted and the top has become golden.

20 Remove from oven, serve it hot.

☞ NOTE

- Sometimes if you over cook the potatoes, the potato skin will become very fragile. Check the potatoes every 15minutes, so it doesn't over cook. Thin potato skin will have same issue, but not because over cook, but because the skin are thin, it can easily break.

- Can add some spicy or fresh chili if desire.

- Can change the fillings.

- You can use tomato instead of potatoes as a vegetable container, but you don't want the center of the tomato as there are too much liquid. You can fill it with anything you like.

Recipe 11

燉飯

Risotto

◆ **材料**

洋菇	4 朵	油	少許
綠花椰菜	⅕ 粒	鮮奶油	1 ～ 2 杯
三色豆（冷凍青豌豆、玉米、紅蘿蔔）	½ 杯	起司	½ ～ 1½ 杯
素料	依個人喜好	鹽巴	少許
米	1 杯	胡椒	少許
高湯	2 ～ 3 杯	帕瑪森起司	依個人喜好

◆ **做法**

1　洗並濾乾米，以同樣動作重複兩～三次。用量米杯裝水後倒入米中，放進煮飯鍋中後，煮至米熟成飯。

2　將洋菇洗淨切片，綠花椰菜以及素料切塊／丁。準備所有食材。

3　準備一個鍋，熱鍋後轉中、小火，加入油，待油熱後加入洋菇以及素料。

4　加入三色豆。

5　加入綠花椰菜，拌炒 3 ～ 5 分鐘。

6　一邊拌炒，一邊慢慢的加入高湯（不可一次加太多），煮至所有蔬菜熟透，略為收汁，留下一點湯汁即可。

7　加入煮好的飯，將兩者混合均勻。

8　加入鮮奶油後煮滾。如果這時醬汁太濃稠，可以再加入一些高湯或是水做調整後煮滾。

9　加入鹽巴、胡椒及起司。

10　待全部材料混合均勻且起司融化，而湯汁呈濃稠狀後，即可享用，享用時可加入帕瑪森起司，風味更佳。

🧑‍🍳 備註

◆ 米可放入煮飯鍋用水或是高湯煮成飯。如果喜歡硬一點的飯，水要放比平常的量再少一些，當你把飯加入醬汁中一起燉煮時，硬度會剛剛好。如果你喜歡軟一點的飯，你煮米的時候可以依正常比例去煮，當你加入醬汁中一起煮的時候，飯會再更軟一些。

◆ 如果你希望正常的煮這道菜，那你將米與蔬菜一起放在鍋內一起拌炒。再炒當中慢慢加入高湯或是鮮奶，每次加入高湯或是鮮奶時都要等收乾再加入，如果沒有適時的加入，鍋內的東西很容易燒焦，持續煮至米熟後，加入調味料，這時你可以加入鮮奶油或是起司，或是兩者都加入也可以。

◆ 將米煮成你喜歡的軟度時，你的食材裡應該仍有許多醬汁、水，若水不夠，可加少許的高湯或水，這樣當你加起司時，食材才會呈濃稠狀。可是如果醬汁不夠，你又加入起司，你的料理會變成塊狀。

◆ 煮好的成品，可倒入烤碗中，並在最上層加入起司，烤箱預熱至 200℃ 後放入烤盤，烤約 5 ～ 10 分鐘，烤至起司融化變成金黃色，即可取出享用。

Risotto

◆ INGREDIENT

Mushrooms	4 whole	Oil	some
Broccoli	⅓ whole	Thick cream or whip cream	1-2 cups
Mix of frozen pea, corn and carrot	½ cup	Cheese	½-1 ½ cups
Soy bean product/meats	as needed	Salt	some
Rice	1 cup	Pepper	some
Vegetable stock/broth	2–3 cups	Parmesan cheese	as needed

◆ METHOD

01 Wash, drain the rice, repeat process 2-3times. Measure the water, and place in the rice, cook it in the rice cooker till it is ready to be used.

02 Wash and cut the mushrooms into slice, and broccoli and soy bean product/meats into chunks/cubes. Prepare all ingredients.

03 Get a pan, turn low or medium heat, add oil, when it is hot, and add the mushrooms, soy bean product/meats.

04 Then add in the mix of frozen pea, corn and carrot.

05 Add in the broccolis. Stir fried it all for 3-5minutes.

06 Add the vegetable stock/broth gradually into the stir fried. Add as you cook in order to make sure all ingredients are cooked soft. Without having too much vegetable stock/broth or juice left over, but it isn't dry.

07 Add in the cook rice. Mix it well.

08 Add in the thick cream or whip cream. Bring it to boil. If it becomes too thick, you can add more vegetable stock/broth or water. Bring it to boil.

09 Add salt, pepper, flavors, and cheese.

10 When it is all well mix, and cheese has melt, the sauce has become thick. It is time to serve. Can add parmesan cheese on top before serving if desire.

🍳 NOTE

• This is the cheating way which give you the same result, which is cook the rice in the rice cooker, but make sure the water or vegetable stock/broth is a little bit less how you cook your rice normally, because you will cook the rice with the sauce, to give it more flavors later. Some people prefer harder rice. So when you cook you will add a bit more less water so later when you cook with the sauce it won't be very soft and make sure when you don't cook it with the sauce, don't cook it too long. If you do like softer dish, than you can cook the rice how you would normally cook it, when you cook it again with the sauce, it will become softer.

• If you would like to cook it proper way, than you cook the raw rice with the vegetables, add vegetable stock/broth or normal milk in gradually until the rice is cook through enough to your liking, make sure every time the vegetable stock/broth or normal milk has been absorbed, you add some more, if you let it dry out, it will burnt the dish, when the rice is cooked you can season it, than you can add the thick cream or whip cream or cheese or both as you desire.

• If there isn't enough sauce, you can add more thick cream or vegetable stock/broth before you add cheese, so when you add cheese, it can become thick and creamy. But if there isn't enough sauce, when you add cheese it will become very lumpy.

• If you like to serve it as a bake dish, than when is it ready, you can dish up on baking dish/container, place more cheese on top evenly, and put it in the oven at 200degree for 5~10mintues, until cheese has melt and becomes golden. Take out and serve hot.

Recipe 12

冷湯麵

Cold Noodle

◆ 材料

新鮮辣椒	1 條	香菇味精	½ 大匙
辣椒碎片或是胡椒子	少許	細麵條	2 ～ 3 片
辣椒粉	少許	香菜	少許
花生碎片	少許	番茄	1 ～ 2 粒
烏醋	2 ～ 3 大匙	小黃瓜	⅓ 條
芝麻香油	½ 大匙	紅蘿蔔	¼ 條
花椒辣油	1 ～ 2 大匙	青、紅椒	各¼ 個
水	4 杯	雪白菇、靈芝菇	各 ½ 個
醬油	2 大匙	蘆筍	½ 把
香菇醬油	少許	豆芽菜	½ 杯
糖	1 ～ 2 大匙	油	1 ～ 2 大匙
鹽巴	少許	高湯	2 杯

◆ 工具

冰塊盒

◆ 做法

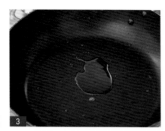
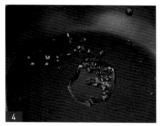

1　洗後將番茄、紅椒、青椒、紅蘿蔔，小黃瓜切至小丁。將新鮮辣椒切小片。香菜剁碎。蘆筍切跟菇類一樣大小（3 ～ 4 公分）。清洗雪白菇、靈芝菇。

2　將高湯煮滾後（煮高湯可以用許多不同蔬菜或是用香菇味精加水煮滾），加入鹽巴調味。放涼後倒至冰塊盒內，放置冷凍庫備用。這也可以前一天先預備好。

3　熱鍋後倒入油，並關至小火。

4　加入辣椒碎片或是胡椒子，翻炒至顏色變深。

5　加入番茄。

6　加入紅蘿蔔、小黃瓜、青椒、紅椒。

7　翻炒 5 ～ 8 分鐘

8　加入水。加入醬油、香菇醬油。

9　加入香菇味精。

10　加入花生碎片，煮至蔬菜變軟。

11　加入雪白菇、靈芝菇，煮 3 ～ 5 分鐘。

12　然後加入蘆筍、豆芽菜，以及所有調味料：芝麻香油、花椒辣油、烏醋、新
　　鮮辣椒、辣椒粉以及糖，煮滾後試味道。確定是自己喜歡的味道後關火放涼。
　　（註：千萬記得，蘆筍不要煮太久，不然顏色不會漂亮。在煮蘆筍時，只要蘆筍的
　　顏色變成漂亮的青綠色就是熟了。如果顏色變成是深綠色，那這樣就是煮太久。）

13　煮一鍋水，煮滾後，加入細麵條，將細麵條煮軟。

14　麵熟後撈起，濾乾，沖冷水。涼後撈起濾乾，然後加到剛剛煮好並放涼的
　　高湯中，再加入之前冰在冷凍庫的高湯冷凍塊。最後上面灑上香菜，即可
　　享用。

Cold Noodle

◆ INGREDIENT

Fresh chili	1 whole
Chili flakes or pepper corns	some
Chili powder	some
Peanut flakes	some
Chinese black vinegar	2-3T
Sesame oil	½T
Peppercorn chili oil	1-2T
Water	4 cups
Soy sauce	2T
Mushroom soy sauce	some
Sugar	1-2T
Salt	some
Dried mushroom stock	½T

Thin noodles	2 -3 piece
Coriander	some
Tomatoes	1-2 whole
Cucumber	⅓ whole
Carrot	¼ whole
Green and red capsicum	¼ whole each color
White beech mushroom, marmoreal mushroom (brown beech mushroom)	½ cup each kind
Asparagus	½ bundle
Bean sprout	½ cup
Oil	1-2T
Vegetable stock/broth	2cups

◆ TOOL

Ice cube tray

◆ METHOD

01 Wash, cut tomatoes, red and green capsicum, carrot, cucumber into small cubes. Cut fresh chili into small slices. Diced coriander finely. Cut asparagus into same size as the mushrooms (3-4cm long). Wash the white beech mushrooms, marmoreal mushroom (brown beech mushroom).

02 Vegetable stock/broth (cook by using different vegetables or use dried mushroom stock with water) bring it boil, season it with salt. Then let it cool. Pour it into the ice cube tray, put in the freezer to be use later. This can be prepare day before.

03 Use a pan or pot, heat it up, add oil, and cook on low heat.

04 Add in the chili flakes or pepper corns. Stir fried it till color is slight change.

05 Add in the tomatoes.

06 Add carrots, cucumbers, green and red capsicums.

07 Stir fried for 5-8minutes.

08 Add in the water. Add the soy sauce, mushroom soy souce.

09 Add in the dried mushroom stock.

10 Add in the peanut flakes. Let it cook till most vegetables are soft.

11 Add in the white beech mushrooms, marmoreal mushroom (brown beech mushroom), and cook for 3-5minutes.

12 Add in the asparagus and bean sprouts and rest of the seasonings: sesame oil, peppercorn chili oil, and Chinese black vinegar, fresh chilies, chili powder and sugar. Bring to boil, taste the flavors. Making sure it is to your liking. Then turn off the heat to let it cool. (Remember when cooking the asparagus don't over cook it; otherwise the color won't be pretty. Try to cook the asparagus till it change color to a lush green, than it is ready. If the asparagus are over cook, it will become a deep green color.)

13 Place water into another pot, bring it to boil, add the thin noodles in, and let it cook till soft.

14 Drain and rinse under cold water. Drain again, and place in the precooked and cooled broth soup. Add in the frozen vegetable stock/broth ice cubes, finish off with coriander on top, it is ready to serve.

🍄 NOTE

• Can add lemon juice, if own desire.

• Can add sesame seeds if desire.

Recipe 13

蕎麥麵

Buckwheat Noodle

◆ 材料

A

海帶芽	½ 杯
甘草	10 片
水	2 ～ 3 杯
香菇味精	½ 小匙
香菇醬油	2 大匙
香菇醬油膏	2 大匙
薑末或薑泥	少許
糖	1 ～ 2 大匙

B

雞蛋	2 ～ 3 顆
小黃瓜	¼ 條
菠菜	1 ～ 2 杯
芝麻	1 ～ 2 大匙
芝麻香油	1 大匙
油	1 大匙
鹽巴	少許
糖	少許
醬油	1 ～ 2 大匙
蕎麥麵	2 ～ 3 捆

◆ 做法

1 準備所有材料。清洗後,將小黃瓜切成細絲。放旁備用。

2 將水加入鍋中,煮滾後,加入菠菜,煮到軟。撈起,沖冷水。濾乾。切成約 3 ～ 4 公分長的小段。

3　將準備好的菠菜放在盤中，加入
　　芝麻香油、鹽巴，最後在上面灑
　　入少許芝麻。

4　再煮一鍋滾水，放入蕎麥麵。煮
　　約 3 〜 5 分鐘，至軟後濾乾，沖
　　冷水並濾乾後放旁備用。

5　將海帶芽放入鍋中，加水，加入
　　甘草、香菇醬油、香菇醬油膏、
　　香菇味精以及糖。

6　用中小火，一邊煮，一邊攪拌。
　　煮至滾後，再滾約 5 分鐘。

7　關火，將甘草撈出。

8　將醬汁倒入一個碗中。

9　再把煮好的海帶芽倒至另外一個
　　容器內，加入薑末或是薑泥，混
　　合均勻，放旁備用。

10　將雞蛋打在一個碗裡，用筷子或
　　是叉子打散，加入醬油、糖，混
　　合均勻。

11　熱鍋後倒入油。

12 油熱後，轉小火，再加入剛打好的雞蛋後，將雞蛋煎成圓形。

13 將雞蛋煎至快要完全成形或是有一點金黃色後翻面。將雙面都煎熟。

14 摺或是捲起煎雞蛋。

15 將雞蛋撈起放至砧板上，切細絲，放旁備用。

16 將所有食材分別裝盤，分成不同區塊，或是將不同食材裝在不同的容器內，即可享用。

🍳 備註

◆ 在享用時，是每一口麵沾取一些醬汁並搭配小菜一起食用。

◆ 醬汁可放置冰箱冷藏約 2 星期。

Buckwheat Noodle

◆ INGREDIENT

A

Dried kelp bud	½ cup
Dried licorice roots	10 slice
Water	2-3 cups
Dried mushroom stock	½t
Mushroom soy sauce	2T
Mushroom soy paste	2T
Grated ginger	some
Sugar	1-2T

B

Eggs	2-3 whole eggs
Cucumber	¼ whole
Spinach	1-2 cups
Sesame seeds	1-2T
Sesame oil	1T
Oil	1T
Salt	some
Sugar	some
Soy sauce	1-2T
Buckwheat noodles	2-3 bundle

◆ METHOD

01 Prepare all ingredients. Wash and cut the cucumber into shreds, set aside ready to use.

02 Add water to a pot, bring it to boil, add spinach into the water, and cook till soft. Drain and rinse under cool water. Drain. Cut into sections around 3-4cm long.

03 Place the spinach on a dish, add on the sesame oil, salt and finish off with some sesame seeds.

04 Bring another pot of water to boil, place the buckwheat noodles in. Let it cook for 3-5minutes, until soft, drain, rinse under cool water. Drain. Set aside ready to use.

05 Place dried kelp bud in a pot, add water, add dried licorice roots, add mushroom soy sauce, mushroom soy paste, dried mushroom stock and sugar.

06 Stir as you cook on low or medium heat. Bring it to boil. Let it cook for 5minutes.

07 Turn off the heat, remove the dried licorice roots.

08 Pour out the sauce into a bowl.

09 Place the dried kelp bud in a separate bowl. Add in the grated ginger. Mix it together, set aside ready to serve.

10 Add the eggs into a bowl; beat it, then add tiny bit of soy sauce and sugar mix well together.

11 Heat a pan, add oil.

12 When oil is hot, cook on low heat. Add in the egg mixture.

13 Cook till eggs are nearly solid/golden, flip over, and cook the same on the other side.

14 Fold or roll it.

15 Place it on a chopping board, cut into thin slice/shreds, set aside ready to use.

16 Plate up together in sections or different plate and serve.

 NOTE

- You dip the noodles into the sauce, each time you eat, with other small dishes.

- The sauce can be store in the fridge for 2 weeks.

大阪燒

Japanese Okonomiyaki

大阪燒

◆ 材料

高麗菜	½ 粒	水或高湯	½ ～ 1 杯
紅蘿蔔	依個人喜好	油	少許
雞蛋	2 ～ 3 顆	鹽巴	少許
麵粉	½ 杯		

配料（依個人喜好加入）：

烤芝麻	少許	鳳梨	少許
海苔絲	少許	美乃滋	少許
玉米粒	少許	醬油膏	少許
番茄	少許	哇沙米	少許

◆ 做法

1 將蔬菜洗淨、高麗菜與紅蘿蔔切成細絲，混合均勻。準備所有的材料，放旁備用。

2 為了幫助高麗菜能夠可以更快的煮熟，可以在高麗菜內加入少許的鹽巴，並蓋上一個盤子，上下搖動 5 分鐘，直到高麗菜摸起來變軟後，用水將鹽巴沖洗掉後濾乾。（註：你可以自己選擇要不要做這個步驟或者跳過這個步驟。）

3 將麵粉倒入碗中，加少許的水。混合均勻至光滑又有一點濃稠度。

4   加入雞蛋並混合均勻。

5   將麵糊與蔬菜絲混合均勻，直到每一個蔬菜都有沾到一些麵糊。如果麵糊太多，做出來的口感會和正常麵糊量的口感不一樣。（註：加入麵糊主要是把蔬菜凝固在一起。）

6   先將小平底鍋熱鍋後，倒入油，轉小火，加入蔬菜糊，厚度約 2 公分，慢慢煎至雙面都呈金黃色即可。如果你是高溫煎的話，中間可能會沒有熟透。

7   當底部已經變成金黃色也有一點酥，翻面，將另外一面也煎成金黃色。

8　裝盤後加入醬油膏（可用刷子刷勻）、美乃滋、哇沙米（如果喜歡的話）。

9　加入烤芝麻、海苔絲、玉米粒、番茄、鳳梨，即可享用。

🧑‍🍳 備註

　　麵糊不可太濃稠、也不可太稀像水，濃稠度應為可在蔬菜上沾薄薄一層即可。

Japanese Okonomiyaki

◆ INGREDIENT

Cabbage	½ whole
Carrot	as needed
Eggs	2-3 whole eggs
Plain flour	½ cup

Water or vegetable stock/broth	½-1 cup
Oil	some
Salt	some

Toppings (Add what you like):

Roasted sesame	some
Seaweed shreds	some
Frozen corns	some
Tomatoes	some

Pineapples	some
Mayonnaise	some
Soy paste	some
Wasabi	some

◆ METHOD

01 Wash and cut the cabbage and carrot into fine shreds, mix it well evenly. Prepare all ingredients, set aside ready to use.

02 To help the cabbages to be able to cook faster, you can add some salt inside the cabbages, cover the top with a plate, and shake it for 5minutes, until the cabbages are soft. Rinse out the salt, and drain. (You can decide to do this step or miss this step. Own choice.)

03 Pour plain flour in a bowl. Add in some water. Mix it till it is smooth and creamy.

04 Add in some eggs into the plain flour mixture. Mix it well.

05 Pour the plain flour mixture into the mix vegetable shreds. Mix it well. Make sure all the vegetables have a thin and light coating of the plain flour mixture. If there are too much of the plain flour mixture, it can still work, but will taste differently to the one with light coating. (Plain flour mixture is to help hold the vegetables together.)

06 Use a small pan, add oil, turn to low heat, add the cabbage mixture in, make sure it's around 2cm thick, and cook it slowly. If you cook it with high heat, the middle won't be cook through.

07 When the bottom has become golden and crispy, flip it over, and cook the other side till the same.

08 Plate up; add soy paste, mayonnaise, wasabi (if desire).

09 Then add roasted sesame, seaweed shreds, frozen corns, tomatoes, pineapples, it is ready to serve.

 NOTE

The plain flour mixture, is not thick or runny like water. The thickness of the mixture should be thick enough to stay on the vegetables.

◆ 材料

米	1 杯	鹹蛋	1 粒
水	3 ～ 4 杯	皮蛋	1 粒
高麗菜	¼ 杯	醬油膏	½ 大匙
三色豆（冷凍青豌豆、玉米、紅蘿蔔）	¼ 杯	鹽巴	少許
任何素料	¼ 杯	糖	少許

◆ 工 具

電鍋

Recipe 15

電鍋煮鹹稀飯

Salty Porridge/Congee

◆ 做法

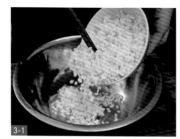

1 洗米後，濾乾；將所有的材料清洗後，切成丁。剝開蛋殼，把鹹蛋與皮蛋切成小丁。

2 熱鍋後加油，轉中小火，將素料炒至金黃色。

3 將洗好的米倒入鍋中，量水後倒入鍋內。

4　加入三色豆。

5　加入高麗菜丁。

6　加入炒好的素料。

7　加入鹹蛋丁、皮蛋丁、醬油膏、鹽巴、糖。

8　用湯匙將所有材料攪拌均勻。

9　放進電鍋中。

10　當電鍋跳起，或是稀飯已經煮好，即可
　　享用（可以依個人喜好加入胡椒或是新鮮
　　辣椒）。

備註

◆ 這個稀飯跟平常煮的稀飯一樣。

◆ 水跟米的比例可依個人喜好做增減。

Salty Porridge/Congee

◆ INGREDIENT

Rice	1 cup
Water	3-4 cups
Cabbage	¼ cup
Mix of frozen pea, corn and carrot	¼ cup
Any type of soybean meat/product	¼ cup
Marinate salty egg	1 whole
Preserved century duck egg	1 whole
Soy paste	½T
Salt	some
Sugar	some

◆ TOOL

Rice cooker

◆ METHOD

01 Wash and drain the rice. Wash and cut all ingredients. Peel the egg shells, cut marinate salty egg and preserved century duck egg into small cubes.

02 Heat a pan, add oil, low-medium heat, stir fried the soy bean product/meats till golden.

03 Pour rice into a container. Add in the measured water.

04 Add in the mix of frozen pea, corn and carrot.

05 Add in the diced cubes of cabbages.

06 Add in the stir fried soy bean products/meats.

07 Add in marinate salty egg and preserved century duck egg, soy paste, salt and sugar.

08 Use a spoon and mix it together.

09 Place in the rice cooker.

10 When it is cooked, it is ready to serve (Can add pepper or fresh chili if desire.).

☞ NOTE

• If you have cook porridge, it is the same way, except with this, you add more ingredient to it.

• The portion of the rice and water can depends on person desire.

素肉鬆

Soy Floss

◆ **材料**

煮熟打碎的大黃豆渣	2～3½ 杯	海苔	少許
芝麻	½～1 杯	糖	1～1½ 大匙
醬油	2～4 大匙		

◆ 做法

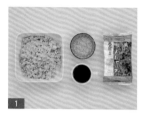   

1 準備所有食材,將海苔剪為絲。

2 小火熱鍋,將煮熟打碎的大黃豆渣倒進鍋中。

3 用筷子翻炒大黃豆渣,直到將大黃豆渣炒乾。大黃豆渣會帶有一點淡淡咖啡色或是比黃色更深一點的顏色。

4 加入醬油,少許的糖,讓味道均衡,不會太鹹。

5 一邊攪拌乾的大黃豆渣,一邊翻炒,一邊慢慢的倒入醬油,讓大黃豆渣吸收醬汁。

6 加入芝麻。繼續翻炒至芝麻顏色改變。關火放涼後,再加入海苔絲。如果鍋中還是熱的,當你加入海苔絲,海苔絲會縮在一起,也會變潮濕。

7 加入海苔絲。將所有食材攪拌均勻。放入一個可以密封的罐子中保存,直到要用的時候再取出。

🧑‍🍳 備註

◆ 如果你希望味道香一點,那當你將大黃豆渣炒乾後,這時你要一邊翻炒一邊慢慢的加入油。直到大黃豆渣吸收所有的油後,再慢慢的加入更多的油。當大黃豆渣不再吸油時,這時候你可以加入少許的醬油,讓大黃豆渣上色。最後加入芝麻跟海苔絲。(註:在這個過程中,你會用較多的油。如果怕太鹹可以加入少許的糖做調味。)

◆ 你也可以加入少許的香菇味精、鹽巴跟胡椒,或是少許的辣椒粉。

◆ 這個食譜可以保存幾個禮拜,不過要確保保存的地方不會潮濕。

◆ 你可以加入其他喜歡的食材。

◆ 如果你覺得翻炒大黃豆渣需要非常多的時間,你可以將大黃豆渣放至烤箱,用小火烤的方式讓大黃豆渣裡面的水分蒸發。要記得需要常常的翻動,才能讓大黃豆渣乾的均勻。

Soy Floss

Crushed cooked soy beans	2-3 ½ cups	Seaweed sheets	some
Sesame seeds	½-1 cup	Sugar	1-1 ½T
Soy sauce	2-4T		

◆ METHOD

01 Prepare all ingredients, and cut the seaweed sheets into shreds.

02 On low heat, heat up a dry pan; pour the crushed cooked soy beans into the pan.

03 Use a chopstick to stir the soy beans, cook till it has dried all the moist in the soy beans. Also the soy beans has a light brown color on it.

04 Add in the soy sauce, a little bit sugar to balance it, if it gets too salty.

05 Continue to stir the soy beans as you add the sauce. Let the soy beans soak up all the sauce.

06 Add the sesame seeds in. Continues to stir fried until the sesame seeds has colored. Turn off the heat, let it cool before adding the seaweed shreds, otherwise the seaweed shreds will become soggy.

07 Add in the seaweed shreds. Combine it all together. And seal in a jar to use when needed.

♟ NOTE

• If you like it more flavor, when the soy beans are all dried, you can add oil slowly, bit by bit, let the soy beans absorb it, until it's not absorbing anymore, and you can add some soy sauce to give it color. Finish off with sesame seeds and seaweed shreds. (During this process, you may need quite a bit of oil, if you feel the flavor is a little bit salty, you can add some sugar to balance the dish.)

• You can also add some dried mushroom stock, salt and pepper, chili flakes into the mixture.

• This can be store for few weeks. But you need to make sure the beans were very dried.

• You can add any dry ingredient to your liking.

• If the stir fried to dry the soy beans takes too long, you can always put it in the oven on low heat to dry it, make sure you turn and mix it during this process so it is evenly dried.

Recipe 17

滷味

Braised Dish

滷味

◆ 材料

馬鈴薯	2～3 粒	肉桂條	1～2 條
紅蘿蔔	1 條	油	少許
高麗菜或大白菜	½ 粒	黑醬油	3～6 大匙
板豆腐或油豆腐	6～10 塊	香菇醬油	2～4 大匙
麵筋球	6～10 塊	香菇醬油膏	3 大匙
海帶	少許	高湯或是水	蓋過食材
黑香菇	5～10 朵	鹽巴	少許
素料	少許	糖	1～3 大匙
甘草	5～10 片	甘蔗汁	2～4 大匙
八角	3～5 顆	五香粉	1～3 小匙

◆ 做法

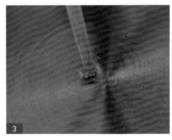

1 將所有食材洗淨、削皮後再將所有蔬菜切塊。準備所有食材。

2 熱鍋後加入油。

3 用筷子來檢查油鍋裡的熱度，如果筷子邊緣有小泡泡，代表油鍋裡的油已經夠熱。

4 轉小火，加入甘草、八角、肉桂條，翻炒至顏色有稍許變咖啡色，或是顏色有變比之前的暗。

5 加入黑醬油、香菇醬油膏、香菇醬油，再翻炒一下。

6 加入少許的高湯或是水，煮滾。

7 加入馬鈴薯、紅蘿蔔、高麗菜或大白菜、黑香菇、素料、板豆腐或油豆腐、海帶、麵筋球，以及五香粉、鹽巴、糖、甘蔗汁調味，煮滾後轉小火，慢慢熬煮20～50分鐘，即可享用。（註：熬煮時間越長會越入味，香味以及味道會更足夠。也可以前一天晚上把湯汁煮好，讓食材泡在湯汁裡，隔天要吃以前把它煮滾，味道會更好。）

🍳 備註

◆當你在炒甘草等食材時，要先加醬油及較稀的醬汁，這樣在炒的過程中，較不易炒焦；若覺得快焦時，關火加入水，把溫度降低即可。（註：若喜好薑，也可加入薑片。）

◆若煮好的食材，有點燒焦味，可加入黑胡椒及芝麻香油，即可蓋過此味道。

◆所有的醬汁及食材，皆可以依個人喜好做調整。

Braised Dish

◆ INGREDIENT

Potatoes	2-3 whole		Oil	some
Carrot	1 whole		Black soy sauce	3-6T
Cabbage or Chinese cabbage	½ whole		Mushroom soy sauce	2-4T
Hard tofu or oily tofu (deep fried ones)	6-10 pieces		Mushroom soy paste	3T
Soy bean balls	6-10 pieces		Vegetable stock/broth or water	just cover all ingredients
Seaweed	some		Salt	some
Chinese black mushrooms	5-10 whole		Sugar	1-3T
Soy bean product/meats	some		Sugar cane juice	2-4T
Dried licorice roots	5-10 pieces		Five spice powder	1-3t
Star anise	3-5 whole stars			
Cinnamon sticks	1-2 whole stick			

◆ METHOD

01 Wash, peel and cut everything into chunks. Prepare all ingredients.

02 Heat a pan, add oil.

03 Use a chopstick to check the heat, if side of the chopstick makes little bubbles, than it is hot enough.

04 On low heat, add dried licorice roots, star anise and cinnamon stick, stir fried it. Until it has become slight brownish, or deeper color.

05 Then add black soy sauce, mushroom soy paste, mushroom soy sauce into it, give another stir.

06 Add some vegetable stock/broth or water. Bring it to boil.

07 Add potatoes, carrots, cabbages or Chinese cabbages and Chinese black mushrooms, soy bean product/meats, hard tofu or oily tofu, seaweed, soy bean balls, Five spice powder and salt, sugar, sugar cane juice into it, bring to boil; let it cook in low heat for 20minutes-50minutes. Ready to serve. (The longer you cook, the more flavors are contain in the ingredients. Sometimes, you can cook night before, bring it all to boil and fully cook, to let cool and soak in the sauce overnight, warm up before serving. Taste even better.)

🍲 NOTE

◆ When you are frying the dried licorice roots mixture, when you add sauce, try to add soy sauce and liquid sauce first, so it doesn't burnt. If you feel it's burning, than quickly, turn off the heat, and add water to bring down the temperature. (Can add ginger slice if desire.)

◆ If your final dish end up with the smell of burnt, add some black pepper and sesame oil, it should cover it.

◆ All the ingredients and sauces, the amount can be alternated to how each desire.

水餃跟煎餃

Dumplings and Fried Dumplings

水餃跟煎餃

◆ 材料

麵粉	2 ～ 3 杯	竹筍	少許
水	2 ～ 2½ 杯	油	2 ～ 4 大匙
冬粉	⅓ 包	香菇醬油	2 ～ 3 大匙
高麗菜	¼ 個	烏醋	少許
紅蘿蔔	¼ 個	香菇味精	少許
黑香菇	3 ～ 6 朵	鹽巴	1 大匙
素料	少許	胡椒	½ 大匙
豆腐	1 ～ 2 塊	水	少許

◆ 工具

擀麵棍

◆ 做法

1　洗、削、切紅蘿蔔，高麗菜切成小丁。
　　豆腐及素料切小丁。將黑香菇泡水後
　　清洗並擠乾、切丁。將冬粉泡軟後切
　　成小段。準備所有材料。

2　將麵粉倒入容器內，加水至麵粉中，
　　在加水時要慢慢的加入，一邊加一邊
　　攪拌。

3　攪拌至麵粉快要成團。

4　將未成形的麵團倒到乾淨的檯面上。

5　用手掌心將麵團搓揉至均勻且光滑。

6　將麵團的表面塗抹一成薄薄的油。放
　　置盤中，蓋起來，醒 15 ～ 20 分鐘。
　　（註：塗油的用意是要讓麵團在醒的時
　　候不會乾掉。）

7　熱鍋後倒入油，開中火待油熱後，加入切好的黑香菇、素料、豆腐一起翻炒至金黃色。在鍋裡還很熱時加入少許的香菇醬油跟烏醋提味。

8　加入紅蘿蔔，竹筍跟高麗菜。

9　將所有食材翻炒均勻，繼續煮約 5 ～ 10 分鐘。

10　加入冬粉、香菇味精、鹽巴、胡椒、少許的水。

11　繼續翻炒至所有的食材都混合均勻，而冬粉也吸飽了醬汁，所有的蔬菜變軟，起鍋呈盤。

12　將麵團分割成小等分約為包餃子的麵團大小，將它搓圓，用掌心將它壓扁，再用擀麵棍將它擀成圓形薄麵皮。

13　將擀好的麵皮放在掌心，挖 1 ～ 2 小匙的餡，放置麵皮的中間。如果你喜歡餡多一點，可以包多一點。但餡若包的不夠多，做出來的水餃會是扁的。

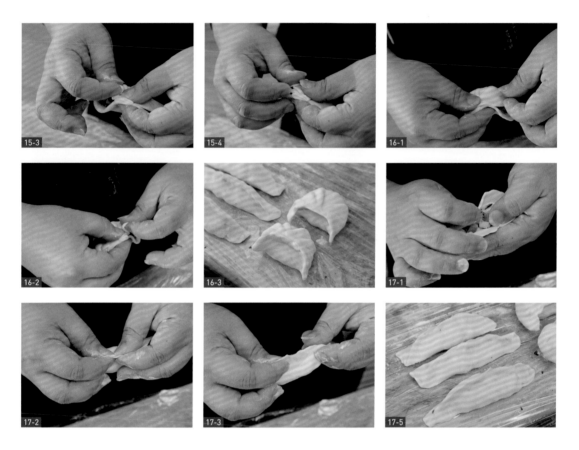

14 將麵皮對折。

15 將麵皮的接合處壓緊後，一隻手捏住
接合處的中心點；另一隻手用大拇指
與食指將旁邊的麵皮往中間拉一點
點，再疊在中心點的上方後摺下去，
壓在一起，持續摺到麵皮尾端。需確
定麵皮都有壓緊，否則在煮的時候接
合處會爆開。

16 然後兩手交換，一隻手拿住麵皮中間，
另外一隻手將麵皮往中間拉，摺的方
式跟之前另外一邊一樣。直到摺完。

17 也可將麵皮擀成橢圓形後對折，為長
條形，做完後像鍋貼。操作時將半圓
的上下壓緊（不用重疊），一直壓緊到
接近麵皮尾端，再將它放在平面上後，
往兩邊壓一下，呈三角形。如果你手
非常的巧，也可以直接用大拇指、食
指跟中指將尾端壓緊在一起。

18 熱鍋，將包好的餃子、鍋貼放在鍋子
裡面。

19 加入少許的油跟一小碗的水。

20 蓋上鍋蓋，讓它煎至水都被吸乾，通
常煎約 5 ～ 15 分鐘。

21 開鍋蓋，這時候你的餃子、鍋貼底層
應該是酥酥的，起鍋，即可享用。

🧑‍🍳 備註

- 麵團越軟，水分越多；麵團越硬，水分越少，做出來的食物口感也會不同。

- 你可以加入少許的鹽巴在麵粉中，讓麵粉有一些味道。

- 你可以在麵團中加入一些蔬菜汁，讓麵皮有其他顏色及味道。若加入蔬菜汁取代水，要把蔬菜汁的水分也算進去，會跟水的份量相同。你可以使用紅蘿蔔汁、紅菜頭汁、菠菜汁等等。

- 你也可以將麵團擀成大大一片，然後用圓形的切割器來切出圓形。然後將剩下的麵團再整理一次，擀開後再切出圓形直到所有麵團都用完。

- 如果你想要增加風味，你可以在內餡中加入少許的香椿提味。

- 你如果喜歡辣，也可以加入一些辣椒粉或是新鮮辣椒。

- 你可以加入任何喜歡的食材在內餡中。

- 配水餃的醬汁：你可以加入醬油膏、醬油、辣椒醬、酸甜醬、甜辣醬，或是自己調配。

  1. 新鮮辣椒、檸檬汁、醬油、少許的糖、芝麻香油跟香菜。

  2. 新鮮辣椒、醬油、烏醋、鹽巴、少許的糖。

  3. 醬油、薑、芝麻香油、少許的糖（新鮮辣椒）。

- 你也可以煮一鍋水，水滾後將水餃加入，煮到水餃浮起來，就熟了，但如果你擔心沒有熟，當水滾了，可以再加一碗冷水，當水再滾第二次時，水餃就熟了。起鍋，瀝水。即可享用。

Dumplings and Fried Dumplings

◆ INGREDIENT

Plain flour	2-3 cups
Water	2-2 ½ cups
Dried green bean noodles/glass noodles	⅓ pack
Cabbage	¼ whole
Carrot	¼ whole
Chinese black mushrooms	3-6 whole
Mixture of soy bean products/meats	some

Tofu	1-2 pieces
Bamboo shoots/slices	some
Oil	2-4T
Mushroom soy sauce	2-3T
Chinese black vinegar	some
Dried mushroom stock	some
Salt	1T
Pepper	½T
Water	some

◆ TOOL

Rolling pin

◆ METHOD

01 Wash, peel and cut the carrot and cabbage into small cubes. Cut the tofu and mixture of soy bean products/meats into small cubes. Soak, wash and squeeze excess water out of the Chinese black mushrooms into small cubes. Soak the dried green bean noodles/glass noodles, until soft, cut it into small short shreds pieces. Prepare all ingredients.

02 Pour plain flour into a mixing bowl, add water to the plain flour, and mix as you add the water.

03 Mix it until it is nearly becoming a dough mixture.

04 Pour the mixture out onto a clean surface.

05 Knead it with the palm of your hand until it is smooth and evenly mixed.

06 Coat the dough outside with a thin layer of oil. Place it onto a plate, cover it with glade wrap and let it rest for 15-20minutes. (Oil is to help to keep the dough hydrated while waiting for it to set).

07 Heat up a pan, add oil, on medium heat, wait for it to get hot, than add the diced Chinese black mushrooms, mixture of soy bean products/meats and tofu. Stir fried it till slight golden. While it is hot, add some mushroom soy sauce and Chinese black vinegar to bring out the flavor.

08 Add the carrots, bamboo shoots and cabbages in.

09 Stir fried till it is all mix and cook it for 5-10minutes.

10 Add in the dried green bean noodles/glass noodles; add in dried mushroom stock, salt, pepper, and some water.

11 Continue to stir fried till it has all combined. And the dried green bean noodles/glass noodles has soaked up all the sauces, the vegetables are nearly soft. Dish up ready to use.

12 Break the big dough into small dough, roll it into a circle, press it flat on a flat surface, use a rolling pin, to roll it out into a circle shaped thin layer sheet.

13 Place the circle dough sheet on your palm; scoop up 1-2t of the filling, place it in the center of the sheet. If you like more filling you can add more, if you have less fillings, it will be a flatter dumplings.

14 Fold it in half.

15 Press the two half dome dough sheets together tightly in the center. With one hand you hold the middle where it is tightly sealed. With the other hand, use your thumb and index finger, pull the top layer dome sheet towards the center and make a fold. Continue to do the same folding technique until the end of the dough sheet. Make sure you seal it tightly so later when you cook it, it won't open up.

16 Than you use the opposite hand to hold the center, and with other hand fold the other side into the center. Continue to fold until it is all finished.

17 You can also make it long ways, where you roll it into an oval shape, you add the filling in the center fold it in half, seal the opening tightly until the end. Than you place it onto a flat surface, press the both side down to make it a flat base, you should get a triangle edge. If you are very good with your hands, you can fold till the end; finish with your thumb and index finger on opposite side of the dome. Use your middle finger to support the bottom. Then seal the ends together by pressing the side down to meet the middle finger.

18 Heat up a pan; place the dumplings in to the pan.

19 Add a bit of oil and a small bowl of water.

20 Cover it with a lid. Let it cook till the water has all be absorbed or dried up which takes 5-15minutes.

21 Remove the lid; you should get a crispy crust on the bottom of the dumplings. Remove from hot pan, ready to serve.

☺ NOTE

- The softer the dough with more water/liquid, the softer finish, the harder dough with less water/liquid, the tougher the finish dumpling it is texture difference.

- You can also add a pinch of salt in the plain flour mixture to give it a little flavor.

- You can also add vegetable juices to give the dumpling plain flour mixture another flavor on the skin, or color. You will still need same amount of the liquid, but it can be divided between the vegetable juice and water. You can use carrot juice to give the skin a orange color skin, or beetroot for a red color, spinach to give a green color look.

- You can roll the dough out in a big flat dough sheet, than use a cup or a round cutter to cut out the circles. Than combine the rest of the dough together, than roll it flat out again, and cut out more circles until the dough has been all used.

- If you want to have more flavors, you can add some toona minced into the vegetables.

- If you like it spicy, you can add chili powder or fresh chili.

- You can add anything vegetables you like into your dumplings.

- Serve with a dipping sauce: you can use a soy paste, soy sauce, chili sauce, sweet and sour sauce, sweet and chili sauce. Or even make your own.

  1. Fresh chili, lemon juice, soy sauce, a little bit sugar, sesame oil and coriander.

  2. Fresh chili, soy sauce, Chinese black vinegar, salt and a bit of sugar.

  3. Soy sauce, ginger, sesame oil, a bit of sugar (fresh chili if like it spicy).

- You can also cook the dumplings in a pot of boil water, when all the dumplings float to the top of the surface, than it is ready to take out. Or if you aren't sure, you can cook the dumplings, bring it to boil, and then add another cup a cold water, let it cook again until it has boiled second time. Than the dumplings are fully cooked. You can drain the water and serve.

中式主菜 CHINESE MAIN COURSE

Recipe 19

豆漿

Homemade Soy Milk

◆ 材料

大黃豆	2 ~ 4 杯
水	8 ~ 12 杯
糖	3 ~ 6 大匙

◆ 工具

過濾袋

大鍋子

果汁機

◆ 做法

1　準備好所有材料。將過濾袋套在一個大鍋子上。放旁備用。

2　洗、泡大黃豆,可在前一天晚上泡或是泡約 7 ~ 10 個小時。

3　將泡好的大黃豆與水放入果汁機內。

4　將食材打到非常細。當然如果喜歡有顆粒就不用打太細,如果到時候你要炒成素肉鬆也比較有口感。

5　把打好的大黃豆倒入一個大鍋子內。
　（註:如果要煮多一點的豆漿,可以多加一點水。）

6　用中小火將大黃豆煮滾,滾後轉小火。煮約 5 ~ 10 分鐘,你也可以煮久一點。在煮的時候要不斷的攪拌,不然大黃豆容易沉澱,沉澱就會燒焦。當然你煮的時間越長,豆漿相對的會比較香也比較濃。

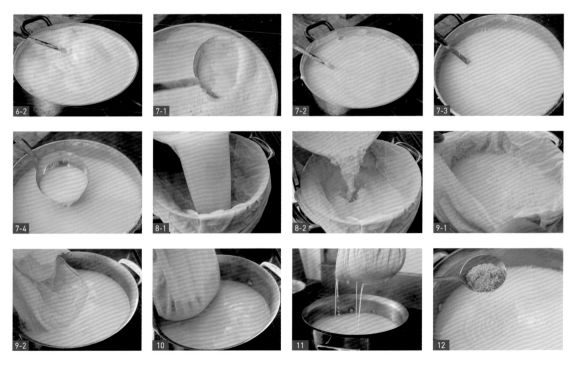

7　在煮的當中，會出現很多泡泡，你可以將泡泡撈出，或是不要管泡泡持續煮，到後面泡泡會慢慢的消失。

8　當大黃豆已經煮好了，將豆漿與豆渣一起倒進過濾袋中。

9　用手拿起過濾袋邊緣，並集中握住。

10　將過濾袋慢慢的往上提起，讓過濾袋中的豆漿慢慢的濾到底下的大鍋子中。

11　將過濾袋拉緊，擠出裡面剩下的豆漿。將豆渣擠越乾越好。（註：你也可以將剛煮好的豆渣再煮一次，但第二次煮的時間較長，大約 15 ～ 20 分鐘，將豆漿味道完全煮出來。你可以將第一次與第二次的豆漿混在一起，這樣味道就會很均勻。）

12　將裝滿豆漿的大鍋子放在爐子上煮滾，加入糖，直到糖融化後試味道。看是不是你喜歡的甜度。確認後關火，放涼。你可以將豆漿冰在冰箱，晚一點再喝，或是可以趁熱的時候喝。

🧑‍🍳 備註

◆ 不要將豆漿放在爐子上煮而沒有攪動。因為豆渣很重會沉到鍋底下，就算水有在滾動，但因豆渣的重量會使它們沉在鍋底，而不會自己翻滾。

◆ 你可加入其他堅果或是豆類一起下去煮，增加不同風味而且也比較健康。

Homemade Soy Milk

◆ INGREDIENT

Soy beans	2-4 cups
Water	8-12 cups
Sugar	3-6T

◆ TOOL

Strong straining bag or nut milk bag
Big pots
Blender

◆ METHOD

01　Prepare all ingredients. Place a strong straining bag over a clean dry pot, set aside to use later.

02　Wash the soy beans, and then soak the soy beans in water overnight, or for 7-10hours.

03　Pour the soy beans and the water into a blender.

04　Blend it till as fine as you like, or as chunky as you prefer. Depends on if you like to further use the ingredient and make it into soy floss. The texture will depend on how fine you blend the soy beans. Of course, the chuncky you left the soy beans, the more crunchy texture you will have.

05　Pour it in a pot. (Can add extra water to cook more quantity.)

06　On low-medium heat, bring the pot of soy beans water to boil. Then turn to low heat. Let it cook it for 5-10minutes or a bit longer if you want. Make sure you keep on stirring the mixture while it is on heat; otherwise it will burn on the bottom. The longer you cook the mixture the stronger soy milk flavor you will get in the end.

07　During this cooking time, there will be lots of bubbles and floss, you can scoop it out, or you can leave it continue to cook, it will disappear later.

08　When it is ready, pour the soy bean mixture into the strong straining bag, along with all the crushed cooked soy beans.

09　Lift the sides of the strong straining bag to one side.

10　Pull the strong straining bag upwards slowly, letting the rest of the soy milk to flow through the strong straining bag.

11　Tighten the bag as the liquid gets less. And squeeze the rest, so the soy beans are as dry as possible. (You can cook the soy beans a second time with some more water, which this time, will need to cook a bit longer than the first, around 10-15minutes, to get as much flavors out as you can, than you combine the first and second batch of the soy milk together to get the same consistency.)

12　Put the pot back on heat, bring it to boil, add sugar, let it dissolve, taste the soy milk to see if it is sweet enough to your liking. Remove from heat, let it cool. Than you can store it in the refrigerator to serve it cold later or you can have it hot.

🍳 NOTE

◆ Do not leave the soy mixture to cook without stirring, it will burn because the soy beans will sink to the bottom, and it is too heavy for it even it's boiling to boil the soy beans apart.

◆ You can add other nuts into the mixture to give extra healthiness and rich flavors.

◆材料

白飯	2～3 碗	高麗菜	⅓ 杯
油	3～5 大匙	鹽巴	½～1 大匙
雞蛋	1～2 顆	胡椒	少許
三色豆（冷凍青豌豆、玉米、紅蘿蔔）	½ 杯	香菇味精	依個人喜好
		醬油	1～2 大匙

◆工具

不沾鍋

Recipe 20

炒飯

Fried Rice

◆ 做法

1　洗乾淨，切高麗菜成小丁，與其他食材差不多大小。準備所有材料。

2　熱鍋後，加入油，待油熱後加入雞蛋。

3　用一個叉子或是一雙筷子，快速的攪拌雞蛋，直到蛋黃和蛋白混合均勻，變成小塊狀。（註：你可以事先將雞蛋拌均勻再放入油鍋拌炒，或是直接將雞蛋打進鍋內，一邊攪拌、一邊炒。）

4　當雞蛋呈現金黃色，取出備用。（註：如果你喜歡比較酥的雞蛋，你可以加入多一點的油，炒至金黃色，這時候若你持續拌炒會出現很多小泡泡；如果你在炒時沒有看到小泡泡代表油不夠多，你可以慢慢加入油，有出現小泡泡是正常的現象。在你關火時小泡泡都會消失。這跟炸蛋的做法很像，只是用的油比較少一點。）

5　加入 3 ～ 5 大匙的油，等油熱後加入高麗菜。

6　翻炒至高麗菜邊緣呈現金黃色。

7 加入三色豆，用中火將所有食材翻炒均勻，但高麗菜偏軟。加入鹽巴、胡椒、香菇味精、醬油，此時口味需調整為較平常再鹹一些，因為待會還會放入白飯，味道就會剛好。

8 加入白飯，用小火翻炒 5 ～ 10 分鐘，直到食材混合均勻而且白飯開始變色或是有一點焦焦的。試味道，如果味道不夠可以再做調整。

9 加入剛炒好的雞蛋，與蔬菜和飯混合均勻後，炒至飯呈現金黃色或是有些燒焦感即可享用。

🍞 備註

◆ 洗米約 2 ～ 3 次後，放入煮飯鍋，米跟水的比例為 1：1。只要是 2 杯米以上，就要多加 1 杯水，例如：2：3 或是 9：10。如果你要飯硬一點，你可以米跟飯的比例一樣。

◆ 如果家裡面沒有煮飯鍋，也可以用平常鍋子煮，一樣可以煮出好吃的飯。將量好的米跟水放入鍋中，開火將水煮滾，然後關小火，煮約 10 ～ 15 分鐘，當你檢查鍋子時，裡面的米開始出現小洞洞，代表水已經被吸收的差不多了。這時蓋上鍋蓋，煮 1 ～ 3 分鐘，關火，不要開蓋，悶約 10 分鐘，讓蒸氣把飯蒸熟，並吸收剩下的水分。

◆ 若用量杯量米，就要用量杯量水，若用碗量米，就要用碗量水。

◆ 炒飯時，若飯是熱的會比較好炒；若要將飯冰在冷凍庫時，要薄薄的一層擺放，之後要使用時，記得要把飯撥開，呈小塊，這樣在炒的時候，較容易跟食材混合，呈現粒粒分明的效果。

◆ 也可以將飯用油事先炒過，提香後再加入醬汁與其他食材。

◆ 炒雞蛋時，用鍋鏟的一角邊畫圓圈，邊炒至雞蛋呈金黃色，或炒至個人喜好的大小即可。

Fried Rice

◆ INGREDIENT

Cooked rice	2-3 bowls
Oil	3-5T
Eggs	1-2 whole eggs
Mix of frozen pea, corn and carrot	½ cup
Cabbage	⅓ cup

Salt	½-1T
Pepper	some
Dried mushroom stock	as needed
Soy sauce	1-2T

◆ TOOL

Non-stick pan

◆ METHOD

01 Wash and cut the cabbage into small cubes, around same size as the other ingredients. Prepare all ingredients.

02 Heat up the pan, add oil, when the oil is hot enough then add the eggs.

03 Use a fork or chopstick, continue stirring the fried eggs until is well mixed and into small pieces. (You can beat the egg mixture before you fried in oil, or you can crack it straightin the pan, and mix it together in the pan while cooking.)

04 When it is golden, take out place aside ready to use. (If you like the eggs more crispy, you can add more oil, let it cook till fully golden, but continue to cook, when you cook it this way, there will be lot of floss and bubbles if you don't see it bubbling, than there isn't enough oil, you can add some more as you go, but it is normal to have lots of floss and bubble, when you turn off the heat, the floss and bubble will all be gone. This is same meaning as deep fried eggs floss, but only in small amount of oil.)

05 Add 3-5T of oil, wait for it to get hot, add in the cabbages.

06 Stir fried it till it is slight golden on the edges.

07 Add the mix of froze pea, corn and carrot, stir fried on medium heat till all combined, and the cabbages are bit soft. Then add salt, pepper, dried mushroom stock, soy sauce, make sure when you combine the flavoring, taste the mixture. The mixture need to be a bit salty, because you will be adding rice into it, which will reduce/dilute the salty flavoring.

08 Add the cooked rice; stir fried on low heat for 5-10 minutes, till it is all combined and the rice has some colors or slightly burnt. Taste it again, making sure it has enough flavors and saltiness. If not enough, this is the time you can add some more flavors.

09 Add the eggs, combine it together, till you see the rice is a bit golden/burnt, and everything is well mixed. It is ready to serve.

🍳 NOTE

◆ You wash the rice, rinse with water 2-3 times, place in rice cooker, ratio of rice and water is 1:1. Or any rice amount that is 2 or more cups, will add one extra cup of water, example 2:3 or 9:10, if you like more harder rice, you can do same amount of rice to same amount of water. Place in rice cooker to cook.

◆ If you don't have a rice cooker at home, you can use a pot with a lid to do the same job, add the amount you like in the pot, bring it to boil, give it a stir, on low heat, let it cook for 10-15minutes, when you see there are lots of little holes in between the rice and the water has reduce. Cover the lid let it cook for 1-3minutes, then turn off the heat, let it sit for 10minutes, when you turn off the heat, do not remove or open the lid, the steam inside after the heat is turned off, will help it to cook through the rice, and the rest of the water will be absorbed into the rice.

◆ If you use a cup to measure the rice, you need to use the same cup to measure the water, if you measure it with a small bowl, you will need to use same bowl to measure water.

◆ It is easier to fried rice, if the rice is hot. Or you can freeze the rice, in a bag, in a thin layer, so when you frying the rice with other ingredient, it will be easier to cook and mix with other ingredient, and the rice won't become sloppy. Make sure you break the rice into small bits or rice is broken into individual rice before you place it in to stir fried.

◆ You can also fry the rice first with oil, to bring out the scent, before adding the sauce and vegetables.

◆ While cooking the eggs, you use one edge of your spatula, turn the eggs clock wise/anti close wise, which ever you are comfortable. Continue this till the eggs are golden, or into the size you like.

◆ 材料

米粉	1 包	鹽巴	1 大匙
高麗菜	½ 顆	糖	½ ～ 1 大匙
紅蘿蔔	1 條	胡椒	½ 大匙
黑香菇	5 ～ 7 朵	醬油	1 ～ 3 大匙
素料	少許	高湯或是水	蓋過食材
油	2 ～ 3 大匙	新鮮辣椒	少許

Recipe 21

炒米粉

Fried MeeHoon

◆ 做法

 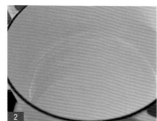

1　將所有蔬菜洗淨，並切成大小差不多的絲。先沖再用冷水泡米粉，把米粉稍微泡軟，濾乾後放旁備用。泡軟的米粉會比較好撥開，也會比較容易炒。

2　熱鍋後，加油，待油熱。

3　放入黑香菇，以及紅蘿蔔。

4　翻炒直至黑香菇呈現金黃色，紅蘿蔔邊緣也稍微有一點金黃色或是變軟。

5　加入素料炒至金黃色。

6　加入高麗菜絲，混合均勻。

7　加入高湯或是水蓋過食材，煮至滾。

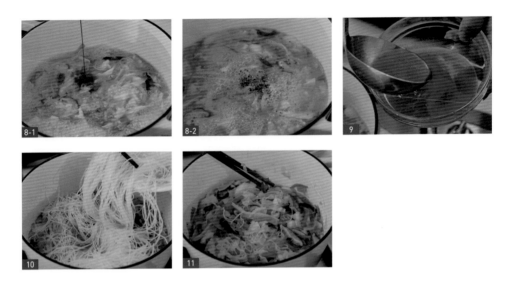

8　加入鹽巴、糖、胡椒、醬油煮滾，試味道。味道要比平常吃的再鹹一點，因為當我們加入米粉時，味道會變淡。

9　舀出一大碗的湯汁，放旁備用。將湯汁舀出後要確保鍋中還剩下足夠的湯汁讓米粉吸收。

10　加入米粉，將米粉與蔬菜混合均勻。要用拉的方式，將蔬菜與米粉一起往上拉起，較容易拌炒，重複這樣的拌炒方式，直到水分被吸乾，再試吃米粉的軟硬度；若覺得米粉不夠軟，可再加入剛舀起的湯汁，繼續拌炒，調整軟硬度。

11　加入新鮮辣椒，混合均勻，即可裝盤享用。（註：如果喜歡辣一點可以早一點加入新鮮辣椒一起拌炒。）

🍳 備註

◆ 黑香菇可以提早幾個小時用冷水浸泡直至軟然後再切，但如果時間不夠可以直接泡 5 ～ 10 分鐘的熱水，清洗後將多餘的水分擠出後再切。之所以用冷水泡是因為要保存黑香菇的香氣，用熱水泡多少會失去黑香菇的香氣。

◆ 可煎雞蛋加入米粉中。

◆ 米粉的味道與材料可依個人喜好做更改。

◆ 之所以將湯汁舀出一大碗是因為如果煮米粉時，湯汁不夠可以再倒回鍋中拌炒，這樣就不會影響米粉的味道。而且若加入自己另外調配的湯汁有時候會很難拌均勻。

◆ 若湯汁已經用完，可用水加入醬油，增加顏色，用小火拌炒，或是加入高湯，會有同樣效果。

◆ 當你浸泡米粉的時間較長，使用的高湯或是水不用很多；但如果浸泡時間較短，相對的，就需要較多的高湯或是水。

◆ 食用前，可依個人喜好加入香菜。

Fried MeeHoon

◆ INGREDIENT

Rice noodles	1 pack	Sugar	½-1T
Cabbage	½ whole	Pepper	½T
Carrot	1 whole	Soy sauce	1-3T
Chinese black Mushrooms	5-7 whole	Vegetable stock/broth or water	
Soy bean product/meats	some		cover all ingredients
Oil	2-3T	Fresh chili	some
Salt	1T		

◆ METHOD

01 Wash, peel and cut all vegetables and Chinese black mushrooms into shreds of even size. Rinse and soak the rice noodles using cold water for few minutes until it is soft to pull it apart slightly. So it will be easier to work with later. Drain, set aside ready to use.

02 Heat up a wok/pot, add oil, and wait till the oil is hot.

03 Add Chinese black mushrooms and carrots.

04 Stir fried it till Chinese black mushrooms become golden, and carrots become slight golden on the edges or have become softer.

05 Add soy bean product/meats stir fried it till golden.

06 Add in the cabbage shreds. Mix it well.

07 Add in the vegetable stock/broth or water, cover all ingredients, and bring it to boil.

08 Add salt, sugar, pepper, soy sauce. Bring it to boil, taste the flavors, make sure it is stronger and a little more saltiness as later when we add the rice noodle, it will reduce the flavors.

09 Than scoop out a big bowl of the sauce/soup, set aside to use if needed. Make sure there are still lots of sauces in the pot for the rice noodles to soak up and cook fully.

10 Add in the rice noodles, try to combine it all together by pulling the rice noodles and vegetables upwards, and then dig it in to the pot again, and pull it up again, repeat till the rice noodles and vegetables are combined, and the sauce has all been absorbed. Taste the rice noodles to see if it is soft enough to your liking, if it's not soft enough, you gradually add the sauce you scoop out earlier back into it, till the rice noodles are soft enough to your liking.

11 Add the fresh chili in, mix it well, dish up, ready to serve. (You can add fresh chili in earlier if you want it to be spicier.)

◆ Chinese black mushrooms, it will be nicer if you can soak in cold water few hours before hand, for it to be soft, wash it, squeeze all the water out, and then cut it. But if you don't have time, then you can use hot water, soak for 5-10minutes, wash and squeeze out the water, then you can cut it. The reason to use the cold water is to keep the flavor in the Chinese black mushrooms, when you use hot water, sometime you loss a bit of the flavor.

◆ You can add fried eggs as well.

◆ The sauce and flavor can change to your own desire.

◆ The reason why you scoop a bowl of the sauce out before you add the rice noodles, is because when you need to add more sauce or liquid, you would have the same sauce as the sauces in the pot. Without needing to add water, which will change the flavors. It is harder to add more flavors into the rice noodles mixture, when there are no more liquid, it will be harder to evenly combine the flavors.

◆ If you do run out of the sauce, and the rice noodles are still hard, you can use this technique to help the rice noodles to cook. Mix a bowl of water with a little bit soy sauce, to give a color and flavor, add gradually into your rice noodles, while cooking in low heat. Or you can just use premade vegetable stock/broth, it works the same.

◆ If you soak the rice noodles longer, the sauce you will need won't be as much, compare to when you soak the rice noodles in shorter time.

◆ Can add coriander on top before serving.

Recipe 22

鹹糕

Chinese Savory Cake

鹹糕

◆ **材料**

粘米粉或是在來米粉
　　　　　　　1 包（大約 450 克）

水　　　　　　　　 10 ～ 11 碗
　（華人常用的普遍喝湯的塑膠碗）

油　　　　　　　　　　　　少許

黑香菇　　　　　　　 3 ～ 6 朵

素料　　　　　　　　　　　少許

菜脯　　　　　　　　　　 2 大匙

青椒　　　　　　　　　　 3 大匙

新鮮辣椒　　　　　　　 ½ 大匙

香菇醬油　　　　　　 4 ～ 6 大匙

黑醬油　　　　　　　 4 ～ 6 大匙

鹽巴　　　　　　　　　　　少許

糖　　　　　　　　　　 1 ～ 2 大匙

胡椒　　　　　　　　 ½ ～ 1 大匙

香菇味精　　　　　　　　　少許

◆ **工具**

電鍋或蒸籠

◆ **做法**

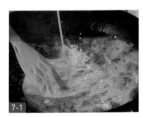

1　洗後將青椒，菜脯切小丁。將黑香菇泡水後清洗，擠出多餘的水分後切小丁。素料切小丁。將新鮮辣椒切末。準備所有材料。

2　熱鍋後加油，待油熱後加入菜脯，翻炒約 5 ～ 8 分鐘，或是等顏色變深，或變金黃色。

3　加入新鮮辣椒跟青椒，稍微調味（如果調過鹹，最後整盤菜會太鹹），繼續翻炒約 3 ～ 5 分鐘。關火，放旁備用。

4　在鍋中加入油，倒入黑香菇丁，翻炒約 3 ～ 5 分鐘，顏色會變成金黃色。

5　加入素料，翻炒至金黃色。

6　加入水，煮滾。加入香菇醬油、黑醬油、鹽巴、糖、胡椒、香菇味精。要調的比平常吃的再鹹一點。

7　在另一個鍋中，加入粘米粉跟水，將兩個材料均勻的攪拌在一起，再加入在滾的水中。一邊慢慢的倒入調好的粘米粉漿，一邊快速的攪拌（最好是用攪拌器攪拌）。如果沒有攪拌均勻，會變成一坨一坨的。

8　當你快要加完所有的粘米粉漿時，鍋中的食材會越來越難攪拌，因為越來越濃稠。所以盡量快速的攪拌均勻，直到所有食材都拌均勻，也呈現濃稠狀，煮滾後關火。如果你覺得很難攪拌，可以先關火，並將粘米粉漿倒入鍋中，與食材攪拌均勻後，再開火煮滾，但在這個過程中，要不斷的攪拌，不然鍋底很容易燒焦。

9　刷一層薄薄的油在容器內。

10　倒入剛煮好的食材並將表面鋪平。

11　把食材放入電鍋中，或是用蒸籠蒸約40～50分鐘。

12　蒸好時，取出，灑上步驟 3 炒好的料。即可享用。

🍳 備註

◆ 如果你想要做蘿蔔糕（芋頭糕），你把蘿蔔（芋頭）洗、削皮、切絲或是刨絲。烹煮的方法一樣，只是在你炒好黑香菇跟素料時，要加入蘿蔔絲（芋頭絲）翻炒約 5～10 分鐘後再加入水，然後繼續剩下的步驟。（註：如果懶的刨絲，你可以將量好的水與蘿蔔一起放入果汁機裡面打，然後再一起放下去煮。當然這樣的做法是比較快，可是做出來的蘿蔔糕比較沒有口感，因為吃不到蘿蔔絲，口感會和芋頭糕一樣。）

◆ 可以加入一些黑醬油，讓鹹糕增加一點顏色，在視覺上不會因成品的顏色過於單調，讓人不想食用。

◆ 在享用時，可以配一些醬油膏、辣椒醬、甜辣醬或是任何自己喜歡的醬。

Chinese Savory Cake

◆ INGREDIENT

Rice flour ___ 1 pack (around 450g)

Water _____ 10-11 bowls
  (Asian standard plastic soup bowl)

Oil _____ some

Chinese black mushrooms
  _____ 3-6 whole

Soy bean product/meats ___ some

Pickled dry radishes _____ 2T

Green capsicums _____ 3T

Fresh chili _____ ½T

Mushroom soy sauce _____ 4-6T

Black soy sauce _____ 4-6T

Salt _____ some

Sugar _____ 1-2T

Pepper _____ ½-1T

Dry mushroom stock _____ some

◆ TOOL

Rice cooker or steamer

◆ METHOD

01 Wash, and cut green capsicums and pickled dry radishes into small fine chunks. Soak, wash and squeeze excess water out of the Chinese black mushrooms, cut it in small chunks same with the soy bean product/meats. Mince diced the fresh chili, prepare all ingredients.

02 Heat up a pan, add oil, and wait for it to get hot, add in the pickled dry radishes. Stir fried it for 5-8minutes, until it slightly change color or till golden.

03 Add in the fresh chili and green capsicums, season it slightly (if you over season it, the finial dish may be very salty), and stir fried for 3-5minutes. Remove from heat, set aside to use later.

04 Add a bit more oil in the hot pan, add in the Chinese black mushrooms. Stir fried it for 3-5minutes. Or till golden.

05 Add in the soy bean product/meats. Stir fried till slight golden on the edges.

06 Pour water in. Bring it to boil. Add mushroom soy sauce, black soy sauce, salt, sugar, pepper, dried mushroom stock, making sure it's a little bit more salty than normally how you would eat.

07 In a bowl mix the rice flour and rest of the water together into a smooth mixture. Add it into the boiling soup; stir as you add the smooth mixture. (It is better to use a whisk.) Otherwise it may get lump and won't have an even consistency.

08 When you get closer to the finish adding the smooth mixture, it will be more thick, and harder to stir, try to mix it together as quick as you can. Until it has all combined, and it is a very thick paste, make sure it is boiling before turning off the heat. But if you find it hard, you can turn off the heat, add rest of the rice flour mixture into the pot, until all is mixed and combined, than you turn on the heat, and cook it till it become boil. During this process you need to continue stirring, so it doesn't burn on the bottom.

09 Brush a thin layer of oil in a container.

10 Pour the mixture into a container. Try to level the top.

11 Place the mixture into a rice cooker, or in a steamer for 40-50minutes.

12 When it is ready, top the stir fried green capsicums filling on the top evenly, it is ready to serve.

⚘ NOTE

- If you like savory turnip cake/radish cake/taro cake. This is how you make it. You wash, peel and shred the white radish; you cook it the same way as the savory cake, after you stir fried the mushrooms and soy bean product/meats, you add the radish in to stir fried for 5-10minutes, add the water, than continue with the recipe. (If you can't be bother to shred the white radish, measure the water, and use most of the water to blend the radish, than it will still have radish flavor, but much quick. But by doing this, you won't get the radish texture when you eating the radish cake, same with taro cake.)

- You can add some dark soy sauce, to give it a little color, otherwise the cake make look a little to white, won't look very nice.

- Can serve with soy paste sauce, chili sauce, sweet and chili sauce, or any sauce to your liking.

Recipe 23

麵腸做法

Gluten Flour Intestinal Roll Method

◆ 材料

麵筋粉（小麥蛋白粉／小麥麵筋粉）........................ 4 杯

麵粉 .... 1 杯

水 .... 5 杯

◆ 工具

刀子　　　夾子

筷子　　　濕布

攪拌器

◆ 做法

1　準備好所有食材，並且把它放在不同的容器內，備用。

2　將一杯麵粉倒入一個容器內。

3　加入水。

4　用攪拌器將麵粉與水攪拌均勻。

5　加入一杯的麵筋粉（小麥蛋白粉）至剛剛拌好的麵糊中，與麵糊攪拌均勻。
　　當麵糊都已經攪拌均勻時，可以再加入下一杯麵筋粉。這時你一樣要攪拌均
　　勻，才可以再加入下一杯麵筋粉。

6　繼續步驟 5 的動作，直到四杯的麵筋粉都已經拌入。這時應該已經形成了一個麵團。可直接用手將麵團揉均勻，搓揉數分鐘。

7　用濕布蓋在容器上，讓麵團醒 8 ～ 10 小時或是放一個晚上。

8　用刀子切一條或是一塊做好的麵筋團出來。

9　用一手的大拇指壓著麵筋團的頂端，另一隻手將麵筋團撐開，一邊拉、一邊撐開麵筋團成薄片狀，大概為快要可以看透麵團一樣的薄度。

10　一隻手拿著筷子，另一隻手拿著麵筋團（慣用的手拿麵筋團），拿筷子的那隻手，用大拇指壓住麵筋團的前端，另外拿麵筋團的手，一邊將麵筋團撐開，一邊拉、一邊撐開麵筋團成薄片狀。

11　拿筷子的手用大拇指與食指開始轉動筷子，將攤開的麵筋片呈 25 度角，且慢慢將麵筋片往下捲，將攤開的麵筋片繞在筷子上，直到繞完為止。

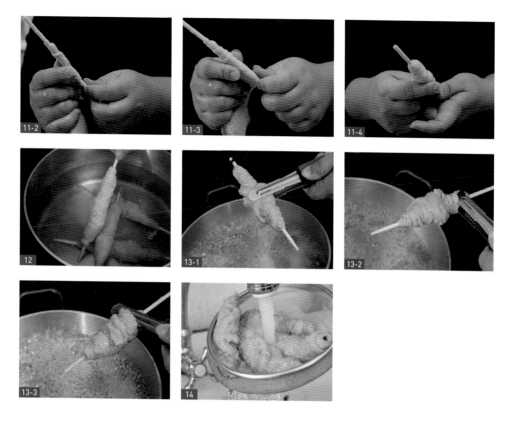

12 備一鍋水，煮滾後，將食材放入滾水中（煮的時候，水不可以沸騰），待麵腸熟時，會浮起來。

13 用夾子將煮好的麵腸取出後，用手拿住筷子的一端，然後用夾子將麵腸推往反方向，將筷子從麵腸裡抽出，再將空心的麵腸放入熱水中，再煮幾分鐘，麵腸會變硬。

14 當麵腸煮好後，取出後沖冷水，即完成。可以將麵腸保存在冷藏或是冷凍中。

👨‍🍳 備註

　　可分成幾個小包裝，放至冷凍保存。如果保存在冷藏，需要 5 天內煮完。

Gluten Flour Intestinal Roll Method

◆ INGREDIENT

Gluten flour (wheat gluten flour)	4 cups
Plain flour	1 cup
Water	5 cups

◆ TOOL

Knife	Tong
Chopsticks	Wet cloth
Whisk	

◆ METHOD

01 Prepare and measure all ingredients. Place in separate container, ready to use.

02 Place 1cup of plain flour in a container.

03 Add the water.

04 Mix it well with a whisk.

05 Add 1cup of gluten flour into the mixture, mixed it well together. When it is all mixed, then you can add another cup of the gluten flour. Again, you need to mix it well before continue to the next cup.

06 Follow same process till all the 4cups has finished and become wet dough, knead it for few minutes till it is well mixed.

07 Use a wet cloth, to cover, over the dough and container, let it rest for 8-10hours or even better overnight.

08 Use a knife; cut it a piece or strips out.

09 Use both hand, hold on either side of the dough, and pull it outwards, trying to stretch it, to check if it is thin and with lots of elasticity. You should be able to pull it till it is very thin, like you can see through the thin layer to the other side.

10 Hold chopstick with one hand, use the other hand (your common used hand) to hold the gluten roll, the hand you hold the chopstick, use thumb to hold down the tip of the strip, while the other hand, pull the strip and spread apart as thin as you can.

11 Turn using your thumb and index finger on the hand with chopstick. As you wrap the strip around the chopstick on the angle of 25degree, slowly wrap it moving down the chopstick, half of the spread dough is over lapping the one before. Till the strip is all finished.

12 Bring a pot of water to boil, then turn down the heat so the water is still, not bubbling while cooking the gluten roll. Then add the finished gluten flour roll that is on chopsticks into the water, when the gluten flour roll is cooked, it will float to the top of the water.

13 Take out the cooked gluten flour roll using a tong, and then remove it from chopstick, by pushing toward one end. Then place the gluten flour intestinal roll back into the hot water (without the chopstick), cook for few more minutes, than the gluten flour intestinal roll will become harder.

14 When it is ready take it out, rinse under cold water. Than it is done. Ready to use or store in freezer/fridge.

🍲 NOTE

Can freeze in small portion in each bag, ready to use. If kept in the fridge, it will need to be cooked within five days.

辣味炒麵腸

Stir Fried Chili Gluten
Flour Intestinal Dish

辣味炒麵腸

◆ 材料

麵腸	5 ～ 10 條	醬油膏	4 ～ 6 大匙
油	少許	香菇醬油	2 ～ 4 大匙
薑末	1 大匙	新鮮辣椒或辣椒醬	1 小匙
烏醋	1 ～ 3 大匙	九層塔	少許
糖	1 ～ 2 大匙		

◆ 做法

1　麵腸用刀子切約 45 度，把麵腸切片或切塊都可以。或是你要用手將麵腸撕成一塊一塊自己喜歡的大小也可以，放置旁邊備用。準備所有食材。

2　熱鍋後，倒入油，等油熱。

3　將切好的麵腸加入鍋中。

4　將麵腸煎至雙面都呈現金黃色。如果喜歡酥一點的，可以煎久一點。可是如果喜歡麵腸軟一點的，煎的時間就不要太長，也不用將麵腸煎得太金黃。當然切麵腸的大小也會影響煎的時間。

5　加入新鮮辣椒或辣椒醬。

6　加入薑末。

7　加入烏醋、糖、香菇醬油、醬油膏，拌炒幾分鐘，直至麵腸吸收了調味料的味道。（註：如果醬汁在你拌均勻前被吸乾，你可以加入少許的水來幫助拌炒，這樣麵腸會均勻的吸收調味料。）

8　加入九層塔。

9　翻炒 2 ～ 3 分鐘，直到九層塔變軟，即可享用。

🍴 備註

◆ 麵腸塊的大小，可依個人喜好；麵腸塊越小，就會越脆。

◆ 也可使用熱油炸麵腸塊，炸至金黃色後取出備用；用油炒薑末至金黃色時，加入烏醋拌炒幾分鐘後，再加入糖、醬油攪拌均勻，最後倒入麵腸塊，混合均勻，即可享用。

◆ 如想要酸一點，烏醋可放置最後和新鮮辣椒或辣椒醬一起加入。

◆ 若要辣一些，可多加一些新鮮辣椒或辣椒醬，或可將薑末與新鮮辣椒或辣椒醬一起用油拌炒。

Stir Fried Chili Gluten Flour Intestinal Dish

◆ INGREDIENT

Gluten flour intestinal roll	5-10 rolls	Soy paste	4-6T
Oil	some	Mushroom soy sauce	2-4T
Grated ginger	1T	Chili or chili sauce	1t
Chinese black vinegar	1-3T	Basil	some
Sugar	1-2T		

◆ METHOD

01 With each gluten flour roll, cut it on 45degree angle chunks or thick slice, or you can even tear it with your hands into pieces of the size you like. Set aside ready to use. Prepare all ingredients.

02 Heat a pan, add oil, and wait till it is hot enough.

03 Then add the gluten chunks/slices.

04 Fried the gluten chunks/slices into crispy golden on both sides. If you like it crispier fried it longer, if you don't like it too hard, then fried it till it has light golden on it. The size of the gluten chunks/slices will also change the frying time.

05 Add the chili or chili sauce.

06 Add grated ginger.

07 Add Chinese black vinegar, sugar, and mushroom soy sauce and soy paste. Stir fried for few minutes till all well mixed, and the sauce has been absorbed. (If the sauce dried up a little too quickly before you can evenly mix it, you can add some water to help it mix well and to be absorbed evenly by the gluten chunks/slices.)

08 Add basil.

09 Stir fried it for 2-3minutes, until basil has softened. Than it is ready to serve.

☺ NOTE

◆ You can divide gluten chunks/slices, into size of own desire. If the size is smaller, it can be more crispy compare to bigger chunks.

◆ You can deep fried the gluten chunks/slices, you just heat the oil, when it's hot enough, add the gluten chunks/slices in, till golden than take it out, set aside to use. Use oil to fried the grated ginger till nearly golden then add Chinese black vinegar to it, stir fried for few minutes, then add sugar, soy sauce, stir it all in. when it is ready, mix it with the gluten chunks/slices, mix well, ready to serve.

◆ If you like it more sour, don't cook the Chinese black vinegar till very last with chili or chili sauce.

◆ If you like it hot and spicy, can add more chili or chili sauce, or fried the chili or chili sauce with grated ginger with oil.

Recipe 25

三杯杏鮑菇

King Mushroom
Chinese Casserole

備註

◆ 在炒醬汁時，可依個人喜好加入胡椒或新鮮辣椒或辣椒醬。

◆ 若想要食材較濃稠，可混合太白粉與水，但需確保鍋內仍有湯汁，若湯汁不夠，可加入高湯或水，需注意在加入太白粉水時，要一邊攪拌一邊加入，直到調整為適當的濃稠度時，再加入九層塔即可享用。

三杯杏鮑菇

◆ 材料

杏鮑菇	4～8朵	醬油	1大匙
九層塔	少許	香菇醬油膏	1大匙
油	2～3大匙	糖	1～2大匙
素沙茶醬	1～2大匙	新鮮辣椒或辣椒醬	少許

◆ 做法

1 將杏鮑菇洗淨後，切滾刀約45度角、3～4公分長的塊狀。切好的杏鮑菇放旁備用，準備所有的材料。

2 熱鍋後加入油，待油熱後加入杏鮑菇。

3 將杏鮑菇炒到金黃色，或是有一些金黃色。

4 加入素沙茶醬。

5 加入香菇醬油膏、醬油、糖，翻炒直到均勻。

6 加入新鮮辣椒或辣椒醬，翻炒幾分鐘後，讓杏鮑菇煮3～5分鐘待杏鮑菇軟。（註：如果杏鮑菇一直沒有軟或是熟。你可以加入少許的水，幫助杏鮑菇煮熟。可是不要加太多水，因為當煮好的時候，大部分的湯汁都會被收乾。）

7 加入九層塔。

8 翻炒1～3分鐘。

9 待九層塔變色或變軟，即可裝盤享用。

King Mushroom Chinese Casserole

◆ INGREDIENT

King oyster mushrooms	4-8 whole
Basil	some
Oil	2-3T
Vegetarian BBQ sauce	1-2T

Soy sauce	1T
Mushroom soy paste	1T
Sugar	1-2T
Chili or chili sauce	some

◆ METHOD

01 Wash and cut (roll cut) the king oyster mushroom in 45degree cubes of size 3-4cm, while turning the king oyster mushrooms. Till all finish. Set aside to rest, prepare all ingredients.

02 Heat pan, add oil, wait till it is hot, than add the king oyster mushrooms in.

03 Stir fried till most of the king oyster mushrooms are golden or golden on some part of the king oyster mushrooms.

04 Add vegetarian BBQ sauce.

05 Add mushroom soy paste, soy sauce and sugar, stir fried it until well mixed.

06 Add chili or chili sauce, stir fried for few minutes and let it cook for 3-5minutes until king oyster mushrooms are soft. (If it isn't soft and fully cooks. You can add some water to help the king oyster mushrooms cook; don't add a lot, because we want the dish to be nearly dry with nice thin coating of sauce.)

07 Add basil.

08 Stir fried for 1-3minutes.

09 Until basil change color or basil has become soft. Plate up and serve.

☞ NOTE

• If you like can add pepper or chili or chili sauce with the sauces, stir fried it together.

• If you like it more thick, you can mix corns starch with water, but you need to make sure there are enough liquid/sauce in the pot, if not enough, add a little bit water or vegetable stock/broth. You pour the corn starch mixture in after the king oyster mushrooms are all cooked, and combine with the sauce, stir the king oyster mushrooms while you add the corn starch water, until the thickness you like, than add basil.

Recipe 26

炒茄子

Stir Fried Eggplant

◆ 材料

茄子	1～2 顆	豆瓣醬	1～1½ 大匙
新鮮辣椒	½ 大匙	糖	1～2 大匙
醬油	1～2 大匙	油	3～5 大匙
醬油膏	1 大匙		

◆ 做法

1　洗淨，切茄子，用滾刀的方式切。新鮮辣椒剁碎。準備所有食材。

2　熱鍋後加油，待油熱後加入茄子。

3　將茄子炸至金黃色或是變軟，取出後濾油。不用一定要放在廚房紙巾上，因為等會炒的時候可以不用再加油。可是如果不希望太油，還是可以用廚房紙巾吸油。

4　熱鍋，加少許的油（只需在鍋上抹上薄薄一層油即可），倒出多餘的油。

5　加入炸好的茄子以及新鮮辣椒，一起翻炒約 3～5 分鐘。

6　加入醬油膏。

7 加入豆瓣醬。

8 加入醬油

9 再翻炒約 3 ～ 5 分鐘。

10 加入糖。

11 將所有食材攪拌均勻。煮到茄子將醬汁都吸收，煮滾。

12 起鍋，即可享用。

🍞 備註

◆ 如果你喜歡比較辣一點，可以加入辣豆瓣醬。

◆ 如果你不喜歡辣，可以不要加入新鮮辣椒，或是將新鮮辣椒切半，將新鮮辣椒裡面的籽去掉，只用新鮮辣椒外皮，增加顏色。

◆ 你也可以加入一些九層塔，添加風味。

◆ 你可以加入少許的高湯或是水，讓這道菜不會太乾；如果是動作較慢的人，可以在煮這道菜時加少許的水，這樣在煮的過程中，有比較多的時間可以做調整。如果湯汁太多，可以加入少許的太白粉，勾芡。

Stir Fried Eggplant

◆ INGREDIENT

Eggplants	1-2 whole	Broad bean paste (bean paste)	1-1 ½T
Fresh chili	½T	Sugar	1-2T
Soy sauce	1-2T	Oil	3-5T
Soy paste	1T		

◆ METHOD

01 Wash and cut the eggplants using roll cut technique. Dice the fresh chili, prepare all ingredients.

02 Heat up a pot of oil, when it is hot, add in the eggplants in.

03 Deep fried it till it become golden or it has becomes soft. Remove from oil, and drain as much oil as possible. Don't have to place on paper towel. Because the excess oil we can use when we stir fried. But if you don't like it too oily, can place on paper towel.

04 Heat up a pan, add oil, enough to coat a light coating on the pan, pour out the excess oil.

05 Add in the deep fried eggplants and fresh chili. Stir fried it for 3-5minutes.

06 Add in the soy paste.

07 Add the broad bean paste.

08 Add in the soy sauce.

09 Stir fried it for 3-5minutes.

10 Add sugar.

11 Combine everything together, and let the eggplants absorbed the sauces. Bring to boil.

12 Than dish up, ready to serve hot.

☞ NOTE

- If you like it spicy, you can always use the broad bean paste with fresh chili in it.
- If you don't like spicy, you can leave the fresh chili out, sometimes you can cut the fresh chili in half, remove the seeds inside, and use just the skin as extra color in the dish.
- You can add basil to give extra flavors.
- You can add some vegetable stock/broth or water, to give it a little sauce so it's not dry when serving. Sometime if you aren't moving fast enough, and you think it's going to burnt, than quickly add few table spoon of broth or water. Give you time to add other ingredients. But if you accidently add too much water. Then you would need to mix some water with corn starch, mix it well, then pour it into the eggplants, to thicken the sauce, but don't overdo the thickening sauce.

◆材料

越南春捲皮	4～8片	薄荷葉	少許
米粉	½包	檸檬汁	少許
紅蘿蔔	⅓條	新鮮辣椒	1大匙
生菜	¼顆	甜醬油	1～3大匙
黑木耳	3～5朵	糖	少許
香菜	少許	水	少許
整顆花生	2-3大匙		

◆工具

深盤子

保鮮膜或烤盤紙

Recipe 27

越南春捲

Vietnamese Spring Roll

◆ 做法

1 將紅蘿蔔、生菜、黑木耳洗、削後切成絲。將香菜與新鮮辣椒切成末。將水加入深盤子中,放旁備用。準備所有食材。

2 煮一鍋水,加入黑木耳煮幾分鐘。

3 濾乾黑木耳,沖冷水,濾乾。放旁備用。

4 同一鍋水,再加入米粉煮幾分鐘,直到米粉變軟。

5 濾乾,沖冷水,再次濾乾。

6 將煮好的米粉切成小段(這樣等一下在包的時候比較好作業),放旁備用。

7 熱一個乾淨的鍋子,放入整顆花生,炒至金黃色,取出放涼後打成小碎塊。放旁備用。或是你可以先把整顆花生弄碎,再加到熱鍋中翻炒直到變色。

8　拿一片乾的越南春捲皮，放入水中稍微
　　泡軟，取出放在乾淨的桌上。越南春捲
　　皮變軟後，需要快速將越南春捲皮取
　　出，不然會軟過頭就很容易破。千萬不
　　要泡過頭，不然會很容易破，不好作業。

9　加入米粉、生菜。

10　加入紅蘿蔔、黑木耳。

11　加入香菜、薄荷葉。

12　加入打碎的花生。（註：不喜歡花生可
　　以不放。）

13　將越南春捲捲皮捲起：先將下面往內摺
　　起（靠近自己的那一邊），包越緊越好，
　　可是要小心不要包破，並確保所有的食
　　材都包在裡面。

14　再將左右兩邊往內折，然後往外推直到
　　底部。沾醬即可享用。

15　包好的越南春捲可以用保鮮膜或是用烤
　　盤紙分隔就不會黏在一起，放在盤子上
　　備用。

16　越南春捲醬：檸檬汁、糖、甜醬油還有
　　新鮮辣椒混合均勻放旁備用。（註：可
　　以加入香菜或是花生增加風味。）

🍳 備註

◆ 越南春捲內的食材，可依個人喜好更改。

◆ 分隔春捲是因為在拿的時候，如果黏在
　一起比較容易破掉。

Vietnamese Spring Roll

◆ INGREDIENT

Vietnamese spring roll wrappers	4-8 piece
Rice noodles	½ pack
Carrot	⅓ whole
Lettuce	¼ whole
Black fungus	3-5 whole
Coriander	some

Peanuts	2-3T
Mint leaf	some
Lemon juice/lime juice	some
Fresh chili	1T
Sweet soy sauce	1-3T
Sugar	some
Water	some

◆ TOOL

Deep plate

Glade wrap or baking paper

◆ METHOD

01 Wash peel, cut carrot, lettuce and black fungus into shreds. Cut coriander and fresh chili into small pieces. Add water into deeper plate, set aside ready to use. Prepare all ingredients.

02 Boil a pot of water, cook black fungus for few minutes.

03 Drain the cooked black fungus, and rinsed under cold water, and drain again. Set aside ready to use.

04 Use same pot of water; cook the rice noodles for few minutes until soft.

05 Drain, raised under cold water, drain again.

06 Cut it into shorter string rice noodles (reason is because if you cut it short, it will be easier to work with). Set aside to use.

07 Heat up a dry pan; add in the peanuts whole, fried till it change color, let it cool and crush it, ready to use. Or you can crush up the whole peanuts into small chunks, add it into a hot dry pan, stir fried it till it change color.

08 Take a piece of Vietnamese spring roll wrapper; soak in the water until it is a little bit softer. Take it out and place onto a flat surface. After the Vietnamese spring roll wrapper is soft, you will need to work fast, otherwise it can break easily. Make sure you don't over soak it; otherwise it can ripe very easy and very hard to work with.

09 Add the rice noodles, lettuces.

10 Add carrots and black fungus.

11 Add corianders and mint leaves.

12 Add crushed peanuts. (If you don't like peanuts, you can miss this step.)

13 Wrap it up by folding the bottom (the side that is closer to you) first towards the center; wrap it as tight as you can without ripping it, making sure all fillings has been wrapped inside.

14 Follow by two either side to the center, when both size has been folded, than roll it outwards to the end of the wrap. Serve with sauce.

15 Place on a plate, make sure each Vietnamese spring roll doesn't touch one and another; you can divide it by using glade wrap or baking paper, so it doesn't stick to each other.

16 The dip: mix lemon juice or lime juice, sugar, sweet soy sauce, and fresh chili, mix well, set aside to use. (Can add some coriander or peanut to give it more flavors.)

☺ NOTE

◆ The filling inside can be change to own desire.

◆ The reason to separate the spring roll is because when it touches one and another, sometime when you pull it apart it will break.

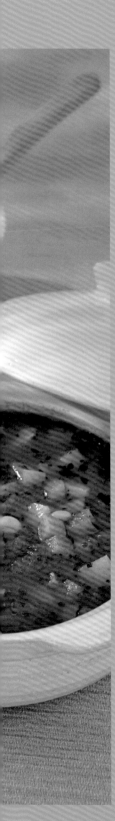

湯

品

VEGETARIAN RECIPES

Soup

◆ 材料

南瓜	2～3 杯
紅蘿蔔	1½ 杯
馬鈴薯	1½ 杯
白花椰菜	1 杯
地瓜	1～1½ 杯
高湯或是水	蓋過蔬菜

鮮奶油	依個人喜好
鹽巴	1～1½ 大匙
胡椒	1 大匙
法式長棍麵包	1 條
奶油	3～6 大匙
義大利香料	1～2 大匙

◆ 工具

錫箔紙

果汁機或馬鈴薯壓泥器

烤盤

Recipe 01

義式濃湯及
法式麵包

Thick Soup with Toasted Bread

◆ 做法

1　將全部材料洗淨、削皮並切片或小塊。

2　把南瓜、紅蘿蔔、馬鈴薯、白花椰菜、地瓜放入高湯內熬煮，煮至蔬菜熟、軟。

3　煮滾後關火，將食材用果汁機打成泥，呈現濃稠狀。

4　加入少許奶油、鹽巴、胡椒、鮮奶油，小火煮滾後即可享用。

5　把切塊的奶油放置碗中。

6　加入義大利香料至奶油中。

199

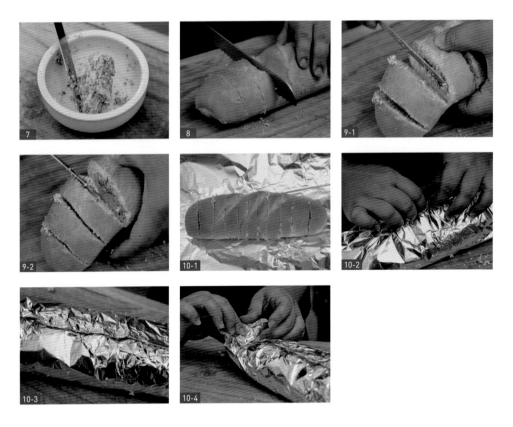

7　充分混合。

8　將法式長棍麵包切 2.5 公分厚,勿完全切斷。

9　把混合好的奶油塗抹在法式長棍麵包夾層中。

10　用錫箔紙將法式長棍麵包包好,不能有空氣進入。

11　烤箱預熱至 200℃後放入已包錫箔紙的法式長棍麵包,烤約 10 分鐘後即可搭配義式濃湯一起享用。

🧑‍🍳 備註

◆ 在做法 5 或是做法 6 的地方,可依個人喜好加入鹽巴跟胡椒。要確保奶油沒有融化,過軟或過硬都會不好操作。

◆ 將所有材料切片的原因在於可縮短熬煮的時間。

◆ 法式長棍麵包也可以搭配義式湯塊的湯頭(位於 P.192 的食譜)。

◆ 加入地瓜可增加湯頭的甜度。

◆ 可依個人喜好選擇是否加入鮮奶油。

Thick Soup with Toasted Bread

◆ INGREDIENT

Pumpkins	2-3 cups
Carrots	1 ½ cups
Potatoes	1 ½ cups
Cauliflower	1 cup
Sweet potatoes	1-1 ½ cups
Vegetable stock/broth or water	cover all vegetables

Thick cream or whip cream	if desire
Salt	1-1 ½T
Pepper	1T
French stick/bread	1 stick
Butter	3-6T
Italian herbs	1-2T

◆ TOOL

Aluminium foil

Blender or electric masher

Baking tray

◆ METHOD

01 Wash all ingredients, peel and cut into slice or small chunks.

02 Place pumpkins, carrots, potatoes, cauliflower, sweet potatoes into the vegetable stock/broth or water, to cook till boil and vegetables are soft.

03 Turn the heat off, use the electric masher to blend all vegetables till well mix, smooth and thick.

04 Add some butter, salt, pepper and thick cream or whip cream, use low heat to cook, bring it to boil. Ready to serve.

05 Mix butter into a bowl.

06 Add Italian herbs into the butter.

07 Mix the butter and Italian herbs until combined.

08 Cut the French stick/bread in 2.5cm thick, but don't cut all the way through, leave bottom part of the French stick/bread still attached.

09 Spread the butter mix between each cut of the French stick/bread.

10 Wrap the French stick/bread with aluminium foil, wrap it tightly.

11 Place in oven 200degree, for 10minutes. Serve it hot with soup.

☺ NOTE

- The butter and Italian herbs can add salt and pepper as desire. Make sure the butter isn't melting, but it is not too hard, because it will be hard to work with.

- Cut all ingredient into slice is because it will be easier to cook.

- French stick/bread can also serve with thick chunky soup (please refer to P.192 for recipe).

- Add sweet potato can give sweetness to the soup.

- You can add thick cream or whip cream if desire.

◆ 材 料

番茄	切丁後 3 杯	西洋芹	¼ 杯
紅蘿蔔	¼ 杯	洋菇	¼ 杯
馬鈴薯	¼ 杯	鹽巴	½ ～ 1 ½ 大匙
玉米粒	¼ 杯	胡椒	1 大匙

Recipe 02

義式湯塊

Thick Chunky Soup

♦ 做法

1 將所有食材洗淨、削皮、切小丁。

2 煮一鍋萬用番茄醬，確保番茄醬不會太濃稠。（註：萬用番茄醬請參考 p.27。）

3 將紅蘿蔔、馬鈴薯、玉米粒、西洋芹、洋菇放入鍋內。

4 煮滾。

5 加入鹽巴、胡椒，用小火煮 30 分鐘後，或是煮至個人喜愛的濃稠度，關火即可享用。

🍳 備註

♦ 若你想要湯多點，可多加一些高湯；若喜歡濃稠一些，高湯加入蓋過食材的量即可。

♦ 可用高湯、義大利香料取代萬用番茄醬。

♦ 可放一些迷你義大利麵一起下去煮。

Thick Chunky Soup

◆ INGREDIENT

Tomatoes	diced 3 cups	Celery	¼ cup
Carrot	¼ cup	Mushroom	¼ cup
Potato	¼ cup	Salt	½-1 ½T
Frozen corns	¼ cup	Pepper	1T

◆ METHOD

01 Wash, peel and cut all ingredients into small cubes.

02 Cook a pot of universal tomato sauce; make sure it is not thick. (Please refer to P.27 for universal tomato sauce recipe.)

03 Place carrots, potatoes, frozen corns, celery, mushrooms into the pot.

04 Bring it to boil.

05 Add salt, pepper cook in low heat for 30minutes. Or until it is thick enough to your liking. Turn off heat. Than ready to serve.

⬠ NOTE

• If you like it with more soup, than you can add more vegetable stock/broth, if like it more thick, than keep it just above ingredient.

• If make by scratch, you can mix vegetable stock/broth with Italian herbs, insteand of using universal tomato sauce.

• Can place small pasta in with the ingredient to cook, if desire.

Recipe 03

羹湯

Chinese Thick Soup

🍞 備註

◆ 蔬菜跟湯的比例可依個人喜好加入水。

◆ 攪拌太白粉時，水量盡量不要太多，因為這樣會把原本湯調好的味道稀釋掉。

◆ 羹湯的味道，應比一般喝湯時的味道再鹹一些，這樣配飯才會剛好。

羹湯

◆ 材料

材料	份量	材料	份量
大白菜	½ ～ 1 粒	高湯或是水	4 ～ 6 杯
紅蘿蔔	1½ 條	太白粉	少許
素料	少許	黑醬油	3 ～ 6 大匙
髮菜	少許	香菇味精	少許
黑香菇	5 ～ 8 朵	烏醋	依個人喜好
黑木耳	5 ～ 8 朵	鹽巴	½ ～ 1 大匙
金針菇	1 把	糖	½ ～ 1 大匙
油	1 ～ 2 大匙	白胡椒	少許

◆ 做法

1 將所有蔬菜洗淨、削皮,將大白菜與紅蘿蔔切成細絲狀。浸泡然後清洗髮菜、黑香菇以及黑木耳後濾乾;將全部的材料切絲,除了髮菜以及金針菇(尾巴要切掉,把它撕成小把的金針菇)。準備所有食材。

2 熱鍋後加入油。

3 油熱後加入黑香菇絲、翻炒至金黃色。

4 加入紅蘿蔔絲,翻炒約 1 ～ 3 分鐘。

5　加入黑木耳，翻炒均勻。

6　加入高湯或是水。

7　加入素料、大白菜、金針菇、髮菜。

8　煮滾。加入黑醬油、香菇味精、烏醋、鹽巴、糖、白胡椒。繼續煮約 5 ～ 10
　　分鐘，直到食材變軟。先試味道，要比平常吃的再鹹一點點，因為當我們加
　　入太白粉時，味道會變淡。

9　將太白粉放置小碗中，加入少許的水混合均勻後，待湯還在滾時，慢慢的加
　　入太白粉水，一邊攪拌，一邊慢慢的加入，如果一次加太多或太快，在湯裡
　　面會結成塊狀，就不會滑順。當你加完太白粉水時，羹湯應該是滑順又濃稠。
　　如果太白粉水不夠。要先關火，將太白粉水調配完後開火，再慢慢的加入，
　　並確認湯有在滾。

10　即可享用。

Chinese Thick Soup

◆ INGREDIENT

Chinese cabbage	½-1 whole		Vegetable stock/broth or water	4-6 cups
Carrots	1 ½ whole		Corn starch	some
Soy bean product/meats	some		Black soy sauce	3-6T
Dried black moss	some		Dried mushroom stock	some
Chinese black mushrooms	5-8 whole		Chinese black vinegar	as needed
Black fungus	5-8 whole		Salt	½-1T
Enokitake mushroom	1 bundle		Sugar	½-1T
Oil	1-2T		White pepper	some

◆ METHOD

01 Wash, peel, and cut Chinese cabbage and carrots into shreds. Soak and wash the dried black moss, Chinese black mushrooms and black fungus. Drain; cut it all into shreds, except for dried black moss and enokitake mushroom (make sure you cut out the bottom part, and pull apart the enokitake mushroom into small bundles). Prepare all ingredients.

02 Heat a pan, add oil.

03 When it's hot, add Chinese black mushrooms, and give it a stir fried till it is golden.

04 Add the carrots, stir fried for 1-3minutes.

05 Add in the black fungus. Stir fried till it is all well mixed.

06 Add in the vegetable stock/broth or water.

07 Add soy bean product/meats, Chinese cabbages and enokitake mushrooms, dried black moss.

08 Bring it to boil. Add black soy sauce, dried mushroom stock, Chinese black vinegar, salt, sugar, white pepper. Let it cook for 5-10minutes, until all ingredients are cook and they are soft. Make sure when you taste the soup, it is slightly salty than how you normally eat. Because you will add corn starch, it will reduce the salty flavor.

09 Mix some corn starch and water in a bowl, mix it well. While the soup is still boiling, add the starch mixture slowly into the soup; stir the soup as you add, so it doesn't become lumpy. It should be smooth and thick when it is finish. If the starch mixture isn't enough, turn the heat off, make some more starch mixture, and then turn the heat back on, add it in, and make sure the soup is boiling.

10 Ready to serve.

☐ NOTE

◆ The ratio of the amount of vegetables and soup, can be alternate to own liking, by changing the amount of water that goes into the dish.

◆ When mixing the corn starch mixture, make sure you don't add to much water, as this can result you changing and dilute the flavor of the soup. This means, you can add more corn starch to small amount of water, which will result better than having more water.

◆ The Chinese thick soup should be a bit more salty than normal soup, as this can be eaten with rice, which will bring the soup just right, when serving with rice.

芋頭米粉湯

Taro Rice Noodle Soup

芋頭米粉湯

◆ 材料

芋頭	1 個	香菜	少許
小白菜	2 個	香菇醬油	2 ～ 4 大匙
紅蘿蔔	⅓ 條	鹽巴	1 大匙
素料	少許	糖	½ ～ 1 大匙
蒟蒻麵捲	1 包	白胡椒	½ 大匙
黑香菇	3 ～ 5 朵	水	6 ～ 9 杯
米粉	1 包	油	1 鍋
芹菜	½ 把		

◆ 工具

果汁機

◆ 做法

1　洗、削、切所有蔬菜。將芋頭用滾刀切塊。將香菜以及芹菜切末。將黑香菇泡、洗，擠出多餘的水分後切絲。將蒟蒻麵捲從包裝內取出，沖水、濾乾。準備所有食材。

2　將一部分切好的芋頭與水倒進果汁機中。

3　將芋頭打成濃漿。

4　熱鍋，加油，待油鍋中的油夠熱後，將切好剩下的少許芋頭放進鍋中炸。將芋頭炸至金黃色，取出，濾油。放旁備用。

5　將素料放進油鍋中炸至金黃色。取出，濾油。放旁備用。

6　將米粉放入溫水中浸泡幾分鐘，直至米粉變軟。

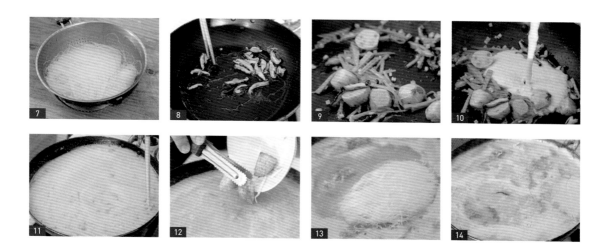

7　濾乾，放旁備用。

8　熱鍋，加油，待油熱後將黑香菇絲倒入鍋中，炒至金黃色。

9　加入紅蘿蔔、芹菜、未炸的素料，翻炒幾分鐘。

10　倒入之前打好的芋頭漿。

11　煮滾，加入香菇醬油、鹽巴、糖、白胡椒。

12　加入炸好的芋頭塊煮至熟後，試味道，湯頭要比較稀，在加入米粉時，才不會因湯被吸乾，變成拌米粉。如果真的太乾，可以加入水，然後再稍微調味。但最好是將湯煮好，量夠再加入米粉與蔬菜。

13　加入米粉與蒟蒻麵捲。

14　加入小白菜。讓所有食材都煮熟、煮滾。不要把米粉煮太軟，不然會變成一坨一坨的。

15　煮好時，用盤子盛裝後，將炸好的素料放在上層，最後加入香菜，即可享用。

🍳 備註

- ◆ 你可以將芋頭事先炸好，保存在冷凍庫，需要時就可以直接使用。
- ◆ 你可以將要打成漿的芋頭炸過，這樣也可以增加風味。
- ◆ 你可以加入任何材料在湯中。
- ◆ 你可以調整湯頭味道，加入少許新鮮辣椒或是醃製品，一起搭配米粉。

- ◆ 你可以藉由水量調整芋頭漿的濃稠度。
- ◆ 可用高湯代替水，跟芋頭一起打，可在湯頭中增加不同風味。
- ◆ 可依個人喜好調整並加入任何素料。
- ◆ 可依個人喜好調整選擇是否加入雞蛋。

Taro Rice Noodle Soup

◆ INGREDIENT

Taro	1 whole
Bok choy (Asian cabbage)	2 whole
Carrot	⅓ whole
Soy bean product/meats	some
Konjac noodle roll	1 pack
Chinese black mushrooms	3-5 whole
Rice noodles	1 pack
Chinese celery	½ bundle
Coriander	some
Mushroom soy sauce	2-4T
Salt	1T
Sugar	½-1T
White pepper	½T
Water	6-9 cups
Oil	a pot

◆ TOOL

Blender

◆ METHOD

01 Wash, peel and cut all vegetable ingredients. Cut the taro in chunks using the roll cut technique. Fine diced/mince coriander and Chinese celery. Soak, wash and squeeze out all excess water from the Chinese black mushrooms and cut into shreds. Remove konjac noodle roll from package, rinse under water, drain. Prepare all ingredients.

02 Add taro chunks and water into the blender.

03 Blend it together into a smooth thick paste.

04 Heat up a pot of oil, let it get hot, add the taro chunks into the oil. Deep fried the taro till golden, remove from oil with a tong, drain oil, set aside to use.

05 Deep fried the soy bean product/meats in the hot oil. When it becomes crispy and brown, remove from oil, and set aside to use later.

06 Soak the rice noodles in warm-hot water for few minutes until it becomes soft.

07 Drain and set aside to use.

08 Heat up a hot pan, add oil, when it is hot, add the Chinese black mushrooms, stir fried till golden.

09 Add in the carrots, Chinese celery, and soy bean product/meats, stir fried it for few minutes.

10 Pour in the blend taro mixture.

11 Bring it all to boil. And add mushroom soy sauce, salt, sugar, white pepper.

12 Add in the deep fried taro and let it cook through. Taste the flavors, making sure it's to your liking. Making sure the soup is very liquid; otherwise when you add the rice noodles later, it will dry up. If that's the case, you can add more water, and season a bit more. Have the soup ready before add the rice noodles and vegetables.

13 Add in the rice noodles and konjac noodle roll.

14 Add in the bok choy. Let everything cook through and bring it to boil. Don't cook the rice noodles too soft, otherwise it will become lumpy.

15 When it is all finish, dish up in a bowl, add the deep fried soy bean product/meats on the top, than finish off with some fresh corianders. Ready to serve hot.

- You can deep fried entire taro beforehand, and place in freezer to store for later use.

- The blend taro can be deep fried as well; it is up to you how you like it.

- You can add any type of ingredients in the soup.

- You can alternate the flavors by adding fresh chili or pickles on the top to serve with the rice noodles.

- You can alternate the thickness by alternating the taro you blend in the water.

- You can also use vegetable stock/broth to blend with the taro instead of using water to give extra richness to the soup.

- You can use all sorts of soy bean product/meats that you are comfortable and familiar with.

- You can add eggs if desire.

點

心

點心 DESSERT

Recipe 01

麵包布丁

Bread Pudding

◆ 材料

雞蛋	2～4 顆
鮮奶	2～3 杯
奶油	1～2 大匙
肉桂粉	½～1 大匙

糖	1～3 大匙
鹽巴	少許
麵包	1～1½ 條／塊

◆ 工具

烤盤或烤碗

◆ 做法

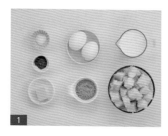

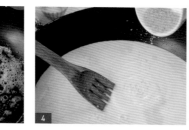

1　將麵包切丁，放旁備用，準備所有食材。

2　熱鍋，加入奶油。

3　待奶油融化後，倒入鮮奶。

4　加入糖。

5　加入鹽巴。

6　加入肉桂粉。

7 攪拌所有的食材,煮到食材都已經融化也混合均勻即可。關火,在加入雞蛋前要讓剛煮好的鮮奶稍微冷卻,不可太燙。不然當加入雞蛋時,雞蛋會被煮成蛋花湯。

8 準備一個碗,打入雞蛋,將雞蛋的蛋白與蛋黃混合均勻。

9 慢慢加入剛放涼的鮮奶,一邊加入一邊攪拌,直到全部都倒入。食材也混合均勻。

10 加入麵包丁。

11 混合均勻。確定每一塊麵包丁都有沾到拌勻的鮮奶與食材。

12 將全部拌好的食材倒入烤盤或是烤碗中。將烤箱預熱 170℃,烤約 30 ～ 40 分鐘或是直到表面呈現金黃色且中心也有熟。用一個乾淨的筷子或是烤肉竹叉,插進中心點看中心是否有烤熟。如果叉子或是筷子拿出來乾淨的,代表食物已經烤熟了。即可享用。

☕ 備註

◆ 如果你的食材是放在烤盤中,而食材沒有很深或很厚,這樣烤的時間會有變動,在烤的時候要注意一下。

◆ 如果食材還沒有烤熟,再讓它多烤一下並反覆檢查。

◆ 如果表面已呈現金黃色,但中心還沒有熟,代表烤箱溫度太高,可是如果中心熟了,可是表面還沒呈現金黃色,則可以將烤箱溫轉高一點,讓表面烤至金黃色即可。

Bread Pudding

◆ INGREDIENT

Eggs	2-4 whole eggs	Sugar	1-3T
Milk	2-3 cups	Salt	some
Butter	1-2T	Bread	1-1 ½ loaf
Cinnamon	½-1T		

◆ TOOL

Baking bowl or tray

◆ METHOD

01 Cut the bread into cubes, set aside ready to use, prepare all ingredients.

02 Heat up a pan, add butter.

03 When butter has melted, pour in the milk.

04 Add in the sugar.

05 Add salt.

06 Then add the cinnamon.

07 Stir till it has all combine, and all the ingredient has melted. Turn off the heat, and cool the mixture before add the eggs. We need to make sure the milk isn't hot enough to cook the eggs; otherwise it will have lumps of cooked eggs.

08 Get a bowl, and add the eggs in, mix the egg white and yolk until combined.

09 Slowly pour it into the milk mixture. Stir till it has fully combined.

10 Add the bread cubes.

11 Mix it evenly. Making sure all the bread has been coated with mixture.

12 Pour into a baking bowl or tray. Then place in the oven at 170degree to cook for 30-40minutes. Or until golden and the inside are cooked. Use a skewer to poke into the middle of the dish to check if the dish is cook thoroughly. The skewer should be clean after taken out from the dish. Ready to serve.

🍳 NOTE

◆ If your ingredient is place in a tray and it isn't very thick, it will change the cooking time. So keep an eye on the dish.

◆ If they cooking time hasn't cook the dish, keep it longer.

◆ If your bread are golden on the top before the inside is cooked, than your temperature in the oven is too high. But if the inside is cooked, but it hasn't color on the top, than you can turn up your temperature, to color the top, till golden.

◆ 材料

紫菜皮 .... 1 包
春捲皮或是餛飩皮 ..... 1 包
芝麻 .... 少許
雞蛋 .... 2 ～ 3 粒
醬油 .... 1 ～ 3 小匙

糖 .... 1 ～ 2 小匙
香菇味精 .... 少許
鹽巴 .... 少許
胡椒 .... 少許
油 .... 1 鍋

◆ 工 具

剪刀
擀麵棍
濕布
廚房紙巾
烤肉刷
有蓋子的容器

紫菜豆皮乾

Seaweed Snack

◆ 做法

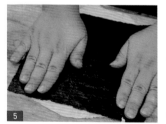

1　將春捲皮或餛飩皮一張張分開（待會操作時不會因黏在一起而不易取用），並用濕布蓋著，較不易乾掉。再另取一條濕布備用。準備所有食材。

2　準備一個碗，打入雞蛋，將雞蛋的蛋白與蛋黃混合均勻，加入醬油、鹽巴、糖、香菇味精、胡椒，混合均勻後，放旁備用。

3　取一個乾的鍋子，熱鍋，將芝麻倒入炒至金黃色，取出，放旁備用。

4　取一張春捲皮或餛飩皮放在桌面，刷一層蛋液。

5　取一張紫菜皮，放在春捲皮或餛飩皮中間，用雙手將兩張皮壓緊。

6　再刷第二層蛋液。

7　平均的灑上芝麻。

8　確定表面幾乎都有灑到芝麻。

9　再放一張春捲皮或餛飩皮在頂層。

10　用擀麵棍將餅皮壓平、固定，且把大部分的氣泡壓出，再用第二條濕布蓋著。放旁備用。

11　當所有的餅皮都做好後，先用剪刀剪成一半或是 ⅓ 的長條狀。

12　再將長條狀剪成寬 2 公分的紫菜皮，放旁備用。

13　熱一鍋油。

14　將紫菜皮放入鍋內，炸至金黃色後撈起、濾乾，放置廚房紙巾上（不要炸成太深的金黃色，因為起鍋時，還有油會持續加熱，這樣會炸過頭）。待冷卻後，就可食用，也可存放至罐子裡，當平常的零嘴食用。

備註

◆ 當你開始組合食材時動作要快，不然你刷過的蛋液容易乾掉，這樣就會沒有黏著性。到開始炸的時候，會容易分離。

◆ 在蛋汁裡面調味是為了在食用時增加風味。如果不想加入鹽巴，可以不用加進去。

◆ 最好用春捲皮做這個食譜，可是如果買不到，餛飩皮也是可以做，只是口感會不太一樣，不過一樣好吃。

◆ 在買春捲皮時不要買到越南春捲皮，如果使用時需要泡水，就代表買錯了。

Seaweed Snack

◆ INGREDIENT

Sushi seaweed wrap _____ 1 pack

Spring roll pastry or wonton
wrappers _____ 1 pack

Sesame seeds _____ some

Eggs _____ 2-3 whole eggs

Soy sauce _____ 1-3t

Sugar _____ 1-2t

Dried mushroom stock
_____ some

Salt _____ some

Pepper _____ some

Oil _____ a pot

◆ TOOL

Scissors

Rolling pin

Wet cloth

Paper towel

Brush

Container with lid

◆ METHOD

01 Separate the spring roll pastry or wonton wrappers, so it will be easier to work with and not stick together; cover it with wet cloth so it doesn't dry up. Wet a second cloth, set aside to use later. Prepare all ingredients.

02 Add eggs into a bowl, add in soy sauce, salt, sugar, dried mushroom stock, pepper into the bowl, then beat the eggs with all the seasoning together until it is well mix. Set aside ready to use.

03 Use a dry clean pan, on low heat; stir fried the sesame seeds, till golden.

04 Get a spring roll pastry or wonton wrappers; lay on a surface, brush a thin layer of egg mixture.

05 Place a sheet of sushi seaweed wrap in the middle of the spring roll pastry or wonton wrappers. Press it down with both hands tightly.

06 Brush another layer of the egg mixture on top of the sushi seaweed wrap.

07 Sprinkle the sesame seeds evenly on the sushi sheet that has been brushed with egg mixture.

08 Make sure you sprinkle a nice decent amount of the sesame seeds.

09 Than place another layer of spring roll pastry or wonton wrappers on the really top.

10 Use a rolling pin, to press the three layers together tightly, try to get as much air bubble out as possible. Set aside cover with the second wet cloth.

11 When all the pastry has been made, use scissor to cut it in half or ⅓.

12 Then cut it into strips of 2cm wide. Set it aside.

13 Heat a pot of oil.

14 Place the sushi seaweed wrap in, deep fried it till golden. Make sure it isn't too golden, because the oil on the sushi seaweed wrap will continue to heat it, if you scoop up too late, it might be dark brown when it has cooled. Scoop up, drain, and let it rest on the paper towel to absorb the excess oil out. Let it cool, serve or store it in a container.

☕ NOTE

- Make sure when you are starting to combine everything, you will need to work very fast, otherwise the eggs will dry on the spring roll pastry or the sushi seaweed wrap, than it won't stick together very well, by the time you deep fried it, it might fall apart.

- The flavors in the egg mixture, is to give the flavors when you eat it, so make sure it isn't too salty, but than it has enough flavors. If you don't want to add salt, you can leave it out.

- The spring roll pastry is better for this recipe, but if you can't buy it, you can use wonton wrappers, it does make a very similar snacks, just texture is a little bit different.

- Make sure your spring roll pastry isn't the Vietnam ones, this one, you don't need to soak in water, if you buy the one that soak in water, than you have purchase the wrong wrap.

Recipe 03

果醬酥皮捲

Jam Pastry Twist

◆ 材料

酥皮	1 包

內餡（可依個人喜好選擇）：

草莓醬	少許
起司	少許
覆盆子醬	少許
巧克力	少許

◆ 工具

果醬刀
刀子
烤盤
烤盤紙
叉子

◆ 做法

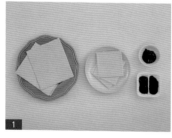
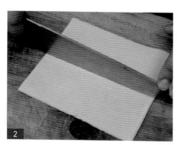

1　準備所有食材。將烤盤紙鋪在烤盤上，放旁備用。

2　取一張酥皮，用刀子切成兩大塊。

3　用果醬刀將果醬均勻的塗滿整個酥皮。

4　用刀子，將塗好果醬的酥皮再切成長條狀。

5　將長條酥皮一個個捲起，變成螺旋狀，即可放上烤盤。（註：如果你懶得捲，可以不用捲，當你塗好果醬後直接放上烤盤。）

6　你也可以將兩種口味捲在一起，兩個單獨捲好時，將兩端的上下疊在一起並捏緊即可放在烤盤上。

7　你也可以切成方形，將巧克力或是起司放在酥皮中心後，對折。

8　用叉子，將邊緣用叉子壓緊，但不可壓太緊以免將邊緣壓斷。

9　將烤箱預熱 200℃，將烤盤放入烤箱烤約 10 ～ 15 分鐘，或烤至金黃色，即可取出享用，也可以放涼再享用。

☕ 備註

◆ 內餡可依個人喜好做更改，但需注意食材要是熟的，因為烤的時間，是要讓酥皮蓬鬆，並非要將食材烤熟。

◆ 可將蛋白打散後，平均塗抹在酥皮上，撒上糖，切條狀後，放入烤箱（也可依個人喜好撒上堅果類於頂端）。

Jam Pastry Twist

◆ INGREDIENT

Puff pastry .... 1 pack

Fillings (own choice, own desire):
Strawberry jam .... some
Cheese .... some
Raspberry jam .... some
Chocolate jam .... some

◆ TOOL

Spread knife
Knife
Baking paper
Baking tray
Fork

◆ METHOD

01 Prepare all ingredients. Lay baking paper on baking tray, set aside ready to use.

02 Get a puff pastry, use a knife and cut it into two sections.

03 Use a spread knife, and spread the jam filling on the whole pastry evenly.

04 Use a knife to cut it into thinner strips.

05 Twist each individual strip, than lay on the baking tray. (If you can't be bothered to twist it, you don't have too. After you spread the jam, you can place it straight on the baking tray.)

06 You can also twist two separate flavors individual strips, when both are twisted, lay it next to each other, then attach the top and bottom together, it become two flavor strips. Than place it on the baking tray.

07 You can cut squares, and add the chocolate jam or cheese in the center, fold it in half.

08 Than use a fork, press down on the edges to seal the edges. But don't press it to hard it went through both layers.

09 Heat the oven to 200degree; place the tray into the oven for 10-15minutes, or until golden. Serve hot or cold.

🍳 NOTE

- Can change the filling to your own desire, but make sure the filling can be eaten; the cooking time is for the puff pastry to puff.
- You can also use egg white, beat it, spread on the puff pastry, sprinkle sugar on the top, cut into strips, and place straight onto the baking tray to bake (can add nut flakes on top if desire).

◆ 材料

麵粉	2 杯
水或鮮奶	2 ～ 2½ 杯
糖	2 ～ 4 大匙
雞蛋	2 ～ 3 顆
油	少許

鮮奶油	1 ～ 2 杯

頂端餡料（依個人喜好加入）：
鮮奶油或是麥芽或是蜂蜜

	少許
當季水果	少許

◆ 工具

攪拌器

過濾網

烤肉刷或廚房紙巾

大湯匙或牛奶壺

Recipe 04

鬆餅

Pancake

◆ 做法

1　洗、切所有的水果。準備所有食材。

2　將鮮奶油放入容器內,加入糖。（註:如果怕太甜,可以放少許的糖。）

3　用攪拌器打發鮮奶油。

4　將鮮奶油打成奶油。當你拿起攪拌器,頂端會呈現出一個尖尖的小端,而且形狀不變。這樣就是好了,放進冰箱,備用。

5　將麵粉加入一個容器中。

6　加入水或是鮮奶至麵粉中。

7　加入糖。

8　加入雞蛋。

9　將所有食材攪拌均勻，呈濃稠狀。

10　拿一個過濾網及一個乾淨的容器，將拌均勻的麵糊倒進過濾網中。

11　用過濾網過濾麵糊，可用攪拌器將麵糊推過過濾網至下面的容器內。用東西蓋住容器上方。讓麵糊醒約15分鐘。

12　熱鍋後，用烤肉刷或廚房紙巾塗一層薄薄的油，轉至小火。

13　用大湯匙或牛奶壺將過濾好的麵糊倒入鍋中，依個人喜好的大小倒入麵糊。

14　當鍋中的麵糊，上面有起泡泡，邊緣處顏色改變，下面呈金黃色，即可翻面。（註：如果你的鬆餅沒有出現很多泡泡，可是邊緣處跟下面都呈現金黃色，這時一樣可以翻面，因為已經煎得差不多了。）

15　將另一面也煎至金黃色，取出後加入自己喜好的食材配料，即可享用。

🍳 備註

◆ 如果你喜歡比較蓬鬆的鬆餅，可以加入½～1小匙的泡打粉跟1～2大匙的油。與麵糊混合均勻。

◆ 若麵粉水非常稀的話，鬆餅就會非常薄；若太濃稠，倒入鍋中時，就無法平均往旁邊散開，所以麵粉水的濃稠度，需比蜂蜜再濃稠一些。

◆ 若麵粉水太稀，可再加入一些麵粉，若太濃稠，則可加入水或鮮奶。

◆ 若不想太甜，糖可依個人需求加在麵糊中，若太甜可加入鹽巴調整；若你要加餡料，要確保麵糊不是太甜，除非是要配水果一起吃即可。

Pancake

◆ INGREDIENT

Plain flour	2 cups
Water or milk	2-2 ½ cups
Sugar	2-4T
Eggs	2-3 whole eggs
Oil	some
Thick cream or whip cream	1-2 cups

Toppings (add how you desire):

Thick cream or whip cream or syrup or honey	some
Mix season fruit	some

◆ TOOL

Whisk

Sift

Brush/paper towel

Spoon/jug

◆ METHOD

01 Wash and cut all the fruits. Prepare all ingredients.

02 Pour the thick cream or whip cream into a container, add in sugar. (Pinch of salt, if you don't like it too sweet.)

03 Use a whisk, and whish the thick cream or whip cream.

04 Whisk until the thick cream or whip cream has becomes thick, and when you lift up the whisk, the cream stay in its shapes. Or you can see some peaks on the tip of the cream. Than it is ready, place it in the fridge for later use.

05 Pour plain flour into a bowl.

06 Add water or milk into the plain flour.

07 Add sugar.

08 Add eggs.

09 Mix it well together so it becomes a thick and smooth batter.

10 Over a clean bowl, use a sift, pour the pancake batter in.

11 Sift the pancake batter through the sift, to remove any lumps. Use a whisk to help, pushing the mixture through. Cover it with a plate and let it rest for 15minutes.

12 Heat a pan, add brush thin lay of oil using a brush or paper towel. Turn to low heat.

13 Pour the pancake batter into the pan using a spoon or jug. Pour it to the size you like.

14 When you see some bubbles on the top of the pancake, and the edge has change color, and the bottom has become golden, you can flip over. (If you don't get lots of bubble, but the side and bottom has change color, you can flip over to cook the other side.)

15 Cook the other side, when both side is golden, dish up, serve hot with own desire toppings.

🍳 NOTE

◆ If you like it a little fluffy and soft pancake, you can add some ½-1t of baking powder and 1-2T of oil into the mixture.

◆ If the plain flour mixture is very running, the pancake will become very thin, if it's too thick, then the pancake won't spread when you add into the pan. So it needs to be a bit thicker than honey is.

◆ If it's too runny, add more plain flour, if it's too thick add more water or milk.

◆ If you don't like it too sweet, make sure you don't add to much sugar in the mixture or you can add a pinch of salt to balance the flavors, also be aware that if you are having topping, make sure your mixture isn't too sweet, unless you are adding fruit instead.

備註

◆ 你可以在煮餡料時加入少許的荳蔻核仁約¼~½小匙，更提味。

◆ 若蘋果沒有軟，可多加些水繼續熬煮，再依個人喜好調整甜度。

◆ 沒有用完的內餡放涼後，可放入保鮮袋，放置冰箱，可保存6個月。

◆ 你也可以用別的餡料代替蘋果。

◆ 你可以用酥皮代替吐司。

Recipe 05

吐司蘋果派

Apple Pie

◆ 材料

吐司	4～8 片	肉桂粉	½～1 大匙
奶油	2～4 大匙	檸檬汁	1 大匙
蘋果	4～6 粒	水	1 杯
糖	2～4 大匙	麵粉	少許

◆ 工具

一鍋油或烤箱
擀麵棍
果醬刀
刀子
烤盤

◆ 做法

1　將蘋果洗淨、削皮、切小丁。準備所有材料。

2　熱鍋，加入奶油，讓奶油融化。

3　加入蘋果丁，翻炒幾分鐘。

4　加入水，蓋過所有的食材。

5　加入糖、肉桂粉、檸檬汁，混合均勻，煮滾。試味道，看是不是自己喜歡的甜度。

6 轉小火慢慢熬煮，至醬汁收乾，且蘋果丁也變軟，入味。

7 將蘋果煮軟，水變少，並依個人喜好調整甜味，再慢慢的加入麵粉，直到蘋果汁變濃稠狀。關火備用。

8 用擀麵棍，將吐司一片片的壓扁。

9 用果醬刀將每一片吐司上下都均勻的塗抹奶油。

10 用刀子將吐司邊切掉。

11 將一片吐司放在乾淨的檯面上，將內餡放在吐司中間。

12 將吐司從靠近自己的那一邊，往外捲。直到內餡都被包在裡面。（如果你麵包變乾了，在捲的時候會容易破掉，所以動作最好快一點。如果太難捲起來，你也可以將它折一半，只要內餡是有包在裡面即可。）你也可以用叉子、刀子、手指頭將開口處封起來。

13 將開口的地方朝下放在烤盤上，如果是對折的，就平放在烤盤上即可。在上方撒少許的肉桂粉跟糖。

14 將烤箱預熱 200℃，放入烤箱烤 5 ～ 10 分鐘至金黃色，或可用油炸至金黃色，即可享用。

Apple Pie

◆ INGREDIENT

Toast	4-8 slice	Cinnamon	½-1T
Butter	2-4T	Lemon juice	1T
Apples	4-6 whole	Water	1 cup
Sugar	2-4T	Plain flour	Some

◆ TOOL

Oven or pot of oil

Rolling pin

Spread knife

Knife

Baking tray

◆ METHOD

01　Peel and cut the apples into small cubes. Prepare all ingredients.

02　Heat a pan, add butter, and let it melt.

03　Add apple cubes. Stir fried it for few minutes.

04　Add water to cover all the apples.

05　Add sugar, cinnamon and lemon juice, mix it well together. Bring it to boil, taste to see if it's the sweetness of your liking.

06　Then let it simmer till the sauce has reduced and thickened also the apples are soft with flavors.

07　Cooking it till the apples are soft, and the water has reduce enough, taste the sauce, make sure the sweetness is your own liking, sprinkle the plain flour in slowly, to thicken the apple juice. Turn off the heat, set aside ready to use.

08　Use a rolling pin, to flatten the toast pieces.

09　Use a spread knife, and evenly spread butter on both sides of the toast.

10　Use a knife, and cut out the sides of the toast.

11　Place a piece of flatten buttered toast on a clean surface; add the apple filling in the middle of the toast.

12　Then roll it, from the side closer to you and roll away from you until all finished, and apple filling is wrap inside the toast. (If your toast is dry, it will break easily, so try to work as quickly as possible before toast gets dried, if it's too hard to roll, than you can fold it in half, as long as the fillings are wrapped inside.) You can even use a fork, knife or your fingers to fully close the edges of the folded toast.

13　Place the open side downwards if it's rolled up, if it's fold in half, and place it on one side of the half. Sprinkle some cinnamon and sugar on the top to finish off.

14　Place in oven to cook at 200degree, for 5-10minutes, till golden, or you can deep fried it in hot oil until golden. Serve it hot.

🍳 NOTE

- You can add some ¼-½t of nut meg in the apple sauce, to give it another extra flavor.

- If the apples are not soft enough, you can add more water to cook it. Taste it, make sure it is sweet to your own tasting.

- The rest of apple pie filling, if there are left overs, you can let it cool, place in a tight container, freeze in the freezer up to 6months. For later use.

- You can use other fillings other than apples.

- You can use puff pastry instead of the bread.

點心 DESSERT

Recipe 06

杏仁千層餅乾

Almond Puff Pastry

◆ 材料

酥皮	6 ~ 10 片
糖	½ 杯
杏仁片	1 ~ 2 杯
蛋	2 ~ 3 顆

◆ 工具

烤肉刷
烤盤
烤盤紙

◆ 做法

1　準備所有的材料。

2　準備一個碗，打入雞蛋，將雞蛋的蛋白與蛋黃混合均勻。放旁備用。

3　將酥皮鋪在一個乾淨的平面上。刷薄薄一層蛋液在酥皮上。

4　確定整片酥皮都有刷到蛋液。

5　平均的撒上少許的糖。

6　將杏仁片均勻的鋪在酥皮上，且不能重疊，否則在烘焙時，杏仁片不會沾黏在酥皮上。

7　用刀子，將酥皮切半。

8　轉約 90 度後，再切成條狀。約 1.5 ～ 2 公分寬。

9　把烤盤紙鋪在烤盤上，然後將每片酥皮條從原本的包裝紙拿起後，放在烤盤上，每片間
　　要保留一些距離，這樣等一下烤的時候它們才有空間膨脹。

10　將烤箱預熱 200℃，放入烤箱烤 10 ～ 15 分鐘，烤至金黃色。從烤箱取出，放涼後保存
　　在一個可以密封的盒子裡面。

🍳 備註

- ◆ 你可以改變裡面的餡料。
- ◆ 要確定表面灑糖的時候有多灑一點，不然
　　會沒有味道。當然如果有細白糖，就不用
　　灑太多，因為細白糖本來就比較甜。
- ◆ 你可以加一些水果乾。

- ◆ 可以加一些新鮮水果，不過要當天吃完比
　　較新鮮。
- ◆ 可以把它改成鹹的，你可以加少許的鹽巴
　　來中和，或是都不要加糖，將上面的餡料
　　改成蔬菜跟起司。

Almond Puff Pastry

◆ INGREDIENT

Puff pastries	6-10 sheets	Almond slices	1-2 cups
Sugar	½ cup	Eggs	2-3 whole eggs

◆ TOOL

Brush

Baking tray

Baking paper

◆ METHOD

01 Prepare all ingredients.

02 Add eggs into a bowl, beat it with chopstick till well mixed. Set aside ready to use.

03 Lay a piece of puff pastry on a flat clean surface. Use a brush, and brush some egg mixture on the puff pastry.

04 Making sure it has covered the whole puff pastry evenly.

05 Sprinkle sugar on top of the puff pastry.

06 Place almond slices on the top of the puff pastry, try to lay it evenly, and not overlapping. If it overlaps, it may not stick to the puff pastry, when it's baking in the oven.

07 Use a knife, and cut the puff pastry in half.

08 Then turn it around 90degree, and cut it into strips around 1.5-2cm thick.

09 Remove it from its original package and place each individual strip on the baking tray, lay it on top of the baking paper, evenly apart with space in between each strip. So later when we bake in the oven, when it expands, it won't stick to the one next to it.

10 Place it in the oven at 200degree, for 10-15minutes, or until it has turn golden. Remove from the oven, let it cool, and store it in a tight seal container.

🍳 NOTE

- You can change to a different topping to your liking.
- Make sure you have enough sugar on top of the puff pastry; otherwise it will be very plain. If you use casted sugar, you don't need to add very much, as the sugar is very sweet.
- You can use dry fruits.
- If you use fresh fruits, than you will need to eat it on the same day.
- Can change this to a savory dish if you desire. Add some salt and less sugar (or no sugar) with topping likes vegetables and cheese.

素食也可以
多滋多味：東西素食大不同。

書　　　名	素食也可以多滋多味： 東西素食大不同
作　　　者	楊雅涵
發　行　人	程顯灝
總　編　輯	盧美娜
主　　　編	譽緻國際美學企業社・莊旻嬑
校稿編輯	譽緻國際美學企業社・劉芸如
美　　　編	譽緻國際美學企業社・羅光宇
封面設計	洪瑞伯
攝　影　師	泰坦攝影事業有限公司

藝文空間	三友藝文複合空間
地　　　址	106台北市大安區安和路二段213號9樓
電　　　話	(02) 2377-1163

發　行　部	侯莉莉
出　版　者	橘子文化事業有限公司
總　代　理	三友圖書有限公司
地　　　址	106台北市安和路2段213號4樓
電　　　話	(02) 2377-4155
傳　　　眞	(02) 2377-4355
E-mail	service@sanyau.com.tw
郵政劃撥	05844889 三友圖書有限公司

總　經　銷	大和書報圖書股份有限公司
地　　　址	新北市新莊區五工五路2號
電　　　話	(02) 8990-2588
傳　　　眞	(02) 2299-7900

初　　版　2017年10月
定　　價　新臺幣420元
ＩＳＢＮ　978-986-364-111-7（平裝）

國家圖書館出版品預行編目 (CIP) 資料

素食也可以多滋多味：東西素食大不同 /
楊雅涵作 .-- 初版 .-- 臺北市：橘子文化，
2017.10
　　面；　公分
　　ISBN 978-986-364-111-7(平裝)

1.素食食譜

427.31　　　　　　　　　　106017186

三友官網

三友 Line@